Lecture Notes in Control and Information Sciences

Edited by M. Thoma and A. Wyner

For information about Vols. 1–61 please contact your bookseller or Springer-Verlag.

Lecture Notes in Control and Information Sciences

Edited by M. Thoma and A. Wyner

128

G. Einarsson, T. Ericson,
I. Ingemarsson, R. Johannesson,
K. Zigangirov, C.-E. Sundberg

Topics in Coding Theory
In honour of Lars H. Zetterberg

 Springer-Verlag Berlin Heidelberg GmbH

Authors

G. Einarsson
Dept. of Teletransmission Theory
University of Lund
P. O. Box 118
S-221 00 Lund
Sweden

T. Ericson
Dept. of Electrical Engineering
University of Linköping
S-581 83 Linköping
Sweden

R. Johannesson
Dept. of Information Theory
University of Lund
P. O. Box 118
S-221 00 Lund
Sweden

I. Ingemarsson
Dept. of Electrical Engineering
University of Linköping
S-581 83 Linköping
Sweden

C.-E. Sundberg
AT & T Bell Laboratories
Room MH 2C483
600 Mountain Avenue
Murray Hill
NJ 07974
USA

K. Zigangirov
Inst. for Problems of Information
Transmission
Ermolova Str. 19
GSP-4, Moscow, 101447
USSR

ISBN 978-3-540-51405-3   ISBN 978-3-540-46208-8 (eBook)
DOI 10.1007/978-3-540-46208-8

Professor Lars H. Zetterberg

Professor Lars H. Zetterberg is the foremost pioneer of information theory in Sweden. He graduated from the Royal Institute of Technology in Stockholm, Sweden in 1949. After employment at the Research Institute of National Defence he completed his degree of licentiate of technology in 1954 and doctor of technology in 1961 also from the Royal Institute of Technology. Zetterberg's thesis for the licentiate degree was titled "A comparison between delta and pulse code modulation".His doctoral thesis covered two topics: "Data transmission over a noisy Gaussian Channel" and " Cyclic codes from irreducible polynomials for correction of multiple errors".

After finishing his doctor's degree Lars Zetterberg was employed by the Swedish Aeroplane Company (SAAB) until 1964 when he was appointed Professor of Telecommunication Theory at the Lund Technical Institute in Sweden. Since 1965 he has had the same position at the Royal Institute of Technology in Stockholm, Sweden.

Zetterberg had a post-doctoral fellowship at the University of Chicago in 1955-56 and was visiting professor at the University of Southern California in 1964. He is internationally recognized and became fellow of IEEE in 1976.

Lars Zetterberg has greatly influenced the research in information theory, communication systems, signal theory, circuit theory and digital technique in Sweden. At present all the professors in Telecommunication Theory and related fields in Sweden are either former students of Zetterberg's or students of former students. He is also recognized as an authority in his field by the telecommunication industry.

Papers by Lars H. Zetterberg

1. A Comparison Between Delta and Pulse Code Modulation.
Ericsson Technics Vol. 11, No. 1, pp 95-154 ,1955

2. Detection of Moving Radar Targets in Clutter.
Information & Control, Vol.1, No. 4, pp 314-33, 1958

3. Data Transmission over a Noisy Gaussian Channel. Trans. of the
Royal Inst. of Technology, Stockholm, No. 184, 1961

4. Study on Data Transmission Under Noise Interference and on Cyclic
Error Correcting Codes. Dissertation, Royal Inst. of Technology,
Stockholm, 1961

5. Cyclic Codes from Irreducible Polynomials for Correction of Multiple
Errors. IRE Trans. on Information Theory, Vol. IT-8, pp 13-20, 1962

6. Signal Detection under Noise Interference in a Game Situation. IRE
Trans. on Information Theory, Vol. IT-8, pp 47-52, 1962

7. Communication with Orthogonal Polyphase Signals over a Noisy
Channel with Doppler Frequency Shift (with Irving S. Reed). IEEE
Trans. Commun. Technol., Vol. COM-12, No. 4, pp 116-18, 1964

8. A Class of Codes for Polyphase Signals on a Bandlimited Gaussian
Channel. IEEE Trans. on Information Theory, Vol. IT-11, No. 3, pp 385-
395, 1965

9. Detection of a Class of Codes and Phase-Modulated Signals. IEEE
Trans. on Information Theory, Vol. IT-12, no. 2, pp 153-161, 1966

10. Analysis of EEG Signals through Estimation of Parameters in a
Linear Model. Proceedings of the annual conference on engineering in
medicine and biology, Vol.10, IEEE, 1968

11. Decision-Adaptive FSK Receivers. Ericsson Tech. (Sweden), Vol. 24,
pp 31-72, 1968

12. Estimation of Parameters for a Linear Difference Equation wih
Application to EEG analysis. Math. Biosciences, No. 5, pp 227 - 75, 1969

13. Signalteori. (With Lars Kristiansson. In swedish.)
Studentlitteraatur, Lund, Sweden, 1970

14. Engineering Aspects of EEG Computer Analysis.(With Kjell
Ahlin.)Eurocon 71 digest, p. 2, IEEE , 1971

15. Application of a Computer-Based Model for EEG Analysis. (With A.
Wennberg.) EEG Clin. Neurophysiol. Vol. 31, No. 4, pp 457-68,
Netherlands, 1971

16. A Predictive Detector and its Applications to EEG. (With Magnus Herolf.) Abstracts, 1973 IEEE International Symposium on Information Theory, p. C1/5

17. Experience with Analysis and Stimulation of EEG Signals with Parametric Description of Spectra. Spike Detection by Computer and by Analogue Equipment. Eds.: P. Kellaway, I. Petersen. Automation of Clin. Electroencephalography, Raven Press, New York, pp 161-201 and 227-234, 1973

18. An Operating System for Computer Analysis of EEG. (With A. Isaksson and A., Wennberg.) Proceedings of the 1st World Conference on Medical Informatics. MEDINFO 74, North-Holland, Netherlands, pp 709-11, 1974

19. Adaptive Delta Modulation with Delayed Decision. (With Jan Uddenfeldt.) IEEE Trans. Commun. Vol. COM-22, No.9, pp 1195-8, 1974

20. Means and Methods for Processing of Physiological Signals. Applications to EEG analysis. European Conference on Electrotechnics. EUROCON'74 Digest. Royal Instn. Engrs., Netherlands, p.E6-1/2, 1974

21. Model of Brain Rhythmic Activity. The Alpha-Rhythm of the Thalamuf. (with F.H. Lopes da Silva, et.al) Kybernetic, No. 15, pp 27-37, 1974

22. Analogue Simulator of EEG Signals Based on Spectral Components. (With Kjell Ahlin.) Medical & Biological Engineering, 1975

23. Means and Methods for Processing of Physiological Signals with Emphasis on EEG Analysis. Advances i Biological and Medical Physics, Vol. 16, Academic Press, 1976

24. Algorithms for Delayed Encoding in Delta Modulation with Speech-like Signals. (With Jan Uddenfeldt.) IEEE Trans. on Commun., Vol. COM-24, pp 652 - 8, No. 6, June 1976

25. Algebraic Treatment of Combined Amplitude and Phase Modualtion by Means of Quaternion Algebra. (With Hugo Brändström.) Abstracts, IEEE International Symposium on Information Theory, p 52, 1976

26. Codes for Combined Phase and Amplitude Modulated Signals in a Four-dimensional Space. (With Hugo Brändström.) IEEE Trans. Commun., Vol. COM-25, No. 9, pp 943-50, 1977

27. Performance of a Model for a Local Neuron Population. (With Lars Kristiansson and Kåre Mossberg.) Biol. Cyber., Vol. 31, pp 15 - 25, No. 1, 1978

28. How to Stimulate Multidisciplinary Research in the Field of Communications Technology. (With Björn Fjaestad.) Proc. ICC'79. p.47.1/1-3, 1979

29. Telecommunication Networks are Developed: Now for Video Services.(In swedish.) Eltek. Aktuell Elektron. Vol. 22, No. 9, pp 46-9, 1979

30. A Microprocessor System for Real Time Analysis of EEG. (With Gunnar Ahlbom and Bengt-Olov Larsson.) Fourth Eropean Conference on Electrotechnics-EUROCON'80, North-Holland, Netherlands; pp 529-33, 1980

31. Computer Analysis of EEG Signals with Parametric Models. (With A. Isaksson and A. Wennerberg.) Proc. IEEE , Vol. 69, No. 4, pp 451-61, 1981

32. Satellite System Provides new Communications Possibilities. (In swedish. With Leif Lundqvist.) Eltek. Aktuell Elektronik. Vol. 25, No. 14, pp 10-17, 1982

33. Interframe DPCM with Adaptive Quantization and Entropy Coding. (With S. Ericsson and H. Brusewitz.) IEEE Trans. Commun. , Vol. COM-30, No. 8, pp 1888-99, 1982

34. Signal Processing for Event Detection. (With L.G. Ahlbom and A. Forsen.) Proceedings of the IEEE International Conference on Acoustics, Speech and Signal Processing, pp 1890-3, 1982

35. Communication Analysis. Technology - but on Human Conditions. (With J.-O. Åkerlund. In swedish.) Tele (Sweden), Vol. 89, No. 1, pp 1-7, 1983

36. Evaluation of COST Project 11 (1972-77) and Project 11bis (1980-1983) on 'Teleinformatics'. (With J. Encarnacao, A. Danthine, E. de Robien, U. Pellegrini, M. Purser and A. Sermet.) Rep. No. EUR 8517 EN, FR Comm. European Communities, Luxemborg, pp xv+71, 1983

37. DPCM Picture Coding with Two-Dimensional Control of Adaptive Quantization. (With S. Ericsson and C. Couturier.) IEEE Trans. Commun., Vol. COM-32, No. 4, pp 457-62, 1984

38. Tree Searching for DPCM Image Coding. (With S. Carlsson.) Proceedings of the International Conference on Communications-ICC 84, Netherlands, Vol. 1, pp 495-501, 1984

39. Event Detection Using Recursively Updated Lattice Filters. (With A. Johansson and G. Ahlbom.) Proceedings of the IEEE International Conference on Acoustics, Speech, and Signal Processing, Vol. 2, pp 632-5, 1985

40. Approximation of Recursively Estimated AR-Parameters for Speech Encoding. (With A. Johansson.) Modelling, Identitication and Robust Control. North-Holland, Netherlands, pp 625-31, 1986

Contents

Detection of On-Off Modulated Optical Signals

Göran Einarsson

Telecommunication Theory, Lund University

Box 118, S-221 00 Lund, Sweden

Abstract—The presentation, which is tutorial in nature, deals with direct detection of noncoherent optical signals. The results are applied to the analysis of optical communication systems using on-off modulation.

The detection situation studied contains optical noise, which exhibits Poisson statistics, together with Gaussian thermal noise. Bounds on and approximations of the detection probabilites for the sum of Poisson and Gaussian stochastic variables are presented and compared with the exact values. The ordinary Chernoff bounds are not tight in this case in contrast to so-called improved Chernoff bounds which are shown to give better accuracy. A convenient way to estimate the detection probabilities is with a saddlepoint approximation which turns out to be accurate to within a few tenths of a percent.

It is shown that system performance can be evaluated with satisfactory accuracy by approximating the compound distribution with either a Poisson or a Gaussian distribution.

1 Introduction

The classical problem in detection theory deals with the detection of signals in additive Gaussian noise which serves as a model for thermal and other noise sources in electronic circuitry.

In optical systems the signal exhibits random fluctuations due to the statistical nature of light as a stream of photons. In contrast to thermal noise the photon flow should be modeled as a Poisson point process. A correct treatment of the detection problem requires calculation of the sum of a Poisson and a Gaussian stochastic variables. The probability density for such a sum is easy to write but it is difficult to handle numerically. In the present paper we show that a convenient way to deal with the situation is to utilize the moment-generating functions of the distributions from which bounds and accurate approximations can be derived.

In almost all textbooks on optical communication the compound distribution is approximated by a Gaussian probability distribution when detection probabilities are calculated. An alternative would be to approximate the compound distribution by a Poisson distribution. These two approaches are shown below to be of about the same complexity and accuracy.

When a more exact evaluation is needed a saddlepoint approximation of the compound density is shown to be a convenient and accurate solution.

The analysis presented applies to ordinary photodiods but can easily be extended to include avalanche photodiods.

2 Photodetectors

The wave nature of light allows it to be modeled as an electromagnetic wave. A complementary description, which is needed in the analysis of optical detection, is the photon model. It specifies light as a stream of minute particles or quanta (photons).

The theory of quantum physics states that the energy of a photon is inversely proportional to the frequency f of the light wave

$$E_p = hf \tag{1}$$

where $h = 6.6261 \cdot 10^{-34} \ W \cdot s^2$ is Planck's constant.

A constant light beam of optical power P corresponds to a stream with an average of P/hf photons per second. For a modulated optical signal with power $P(t)$ the photon intensity varies with time

$$\gamma(t) = \frac{P(t)}{hf} \ \text{photons/sec.} \tag{2}$$

A conventional optical receiver contains as its first element a photodetector which converts the incoming optical signal into an electrical signal. The receiver performs signal processing operations, such as amplification and filtering, on the electrical signal to arrive at a decision about the information conveyed by the optical signal.

Most photo detectors are based on the principle of optical absorption. An incoming photon excites an electric charge carrier (a hole or an electron) moving it from the valence band to the conduction band. A flow of incoming photons generates a current of photoelectrons at the output of the photodetector.

For a modulated optical signal with power $P(t)$ the electron density from the photodetector is

$$\gamma(t) = \frac{\eta P(t)}{hf} \ \text{electrons/sec.} \tag{3}$$

where η is the quantum efficiency of the device. The power $P(t)$ in (3) is the instantaneous intensity of the optical signal. It is proportional to the square of the electromagnetic field quantities in the wave description.

The photons and the photoelectrons appear at random points in time and the stream of electrons from the photodetector is a random process. It can be shown theoretically, and it has been verified experimentally, that a valid statistical model is the Poisson process [10].

3 The Poisson process

Consider a sequence of photoelectrons at a fixed instant of time. The time positions of the electrons is a realization of a Poisson (point) process [2]. The intensity $\gamma(t)$ is a measure of the rate at which events occur. The average number of events in a time interval of length T starting at an arbitrary time t_1 is

$$E\{N(T)\} = \int_{t_1}^{t_1+T} \gamma(t)dt = m \tag{4}$$

The number $N(T)$ of events in the time interval is a random variable with a Poisson distribution

$$Pr(N(T) = n) = \frac{m^n e^{-m}}{n!} \tag{5}$$

The variance of $N(T)$ is [2]

$$E\{(N(T) - m)^2\} = m \tag{6}$$

A property which is useful in the analysis is that the times of occurance of Poisson events in any fixed time interval are independent of each other if the order of the occurrences is disregarded.

3.1 Shot noise

The Poisson process is a point process generating a sequence of random time events representing the times of occurrence of the photoelectrons. A stream of electrons constitutes an electric current. Each electron induces a short current pulse $g(t)$ in the electric circuit and the output electric current from the photodetector is

$$i(t) = \sum_{k=-\infty}^{\infty} a_k g(t - t_k) \tag{7}$$

The current pulse $g(t)$ has an area equal to the charge of an electron

$$\int_{-\infty}^{\infty} g(t)dt = q \tag{8}$$

The current $i(t)$ is a stochastic process generated by the Poisson events t_k. Such a process is called shot noise. The parameter a_k can be a constant, as it is for ordinary photodetectors, or it can be a random variable, as in the case of an avalanche photodiode where a multiplicative process is involved in the production of photoelectrons.

The statistical properties of stationary shot noise $i(t)$ with $\gamma(t) = \gamma$ are summarized below. For further details see e. g. [8]

The average of the current $i(t)$ is

$$i_0 = E\{i(t)\} = E\{a_k\}\,\gamma\,q \tag{9}$$

The power spectral density is

$$R(f) = E\{a_k^2\}\,\gamma\,|G(f)|^2 + i_0^2\,\delta(f) \tag{10}$$

where $G(f)$ is the Fourier transform of $g(t)$

$$G(f) = \int_{-\infty}^{\infty} g(t)e^{-j2\pi ft}dt \tag{11}$$

In practice the time duration of the pulses $g(t)$ is often small. Approximation with a dirac function $g(t) = q\delta(t)$ gives $G(f) = q$ and results in a constant spectral density, i.e. white noise

$$R(f) = E\{a_k^2\}\,\gamma\,q^2 + i_0^2\,\delta(f) \tag{12}$$

If such a white shot noise process is filtered by a bandpass filter with bandwidth B the direct current component is eliminated and the output is a noise signal with variance

$$\sigma_B^2 = 2E\{a_k^2\}\,\gamma\,q^2 B \tag{13}$$

For the special case when $a_k = 1$, corresponding to a photodiode without avalanche gain, σ_B^2 becomes

$$\sigma_B^2 = 2\gamma\,q^2\,B \tag{14}$$

From (13) and (14) it follows that the photodetector shot noise process is signal-dependent. The noise variance depends on the optical signal power. Light of high intensity γ generates a large average output current i_0 but also noise of large variance.

4 Ideal receiver. The quantum limit

We consider binary optical transmission with perfect on-off modulation. The binary symbol "one" is represented by an optical pulse of arbitrary form $p(t)$ confined in the signaling time interval T as shown in Fig. 1. A binary "zero" is indicated by the absence of any optical signal in the corresponding time interval.

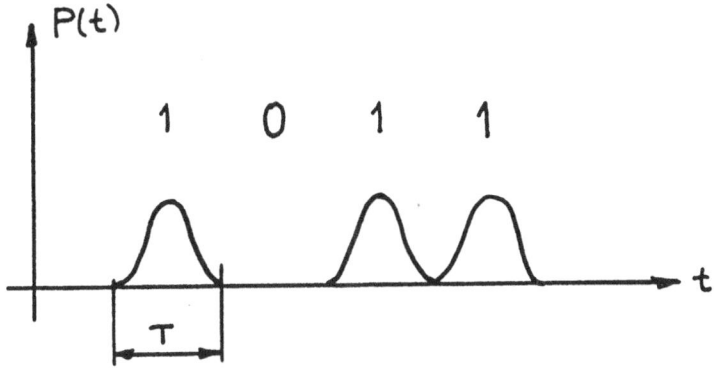

Figure 1:
Binary on-off signaling.

The receiver is assumed to be ideal in the sense that it is capable of detecting every single photon arriving at its input. Such a receiver could in principle be realized as a photodetector with quantum efficiency equal to unity followed by a sensitive electronic counter recording the number of photoelectrons produced by the detector. See Fig. 2.

The number of photons in the pulse $p(t)$ is a random variable. As shown in section 2 it has a Poisson distribution with a mean value obtained from (3) and (4)

$$m_1 = \frac{1}{hf} \int_0^T p(t)dt \qquad (15)$$

For the system to operate satisfactorily we require that the bit error probability should be at most 10^{-9}.

The receiver observes the number of photons in consecutive time intervals. For those intervals where no photons are detected it assumes that a binary zero was transmitted. The only way an error can occur is when a one was transmitted but no photons are detected. With the assumption that the binary symbols are equally likely, which means that they both occur with probability $P = 1/2$, the bit error probability is, see (5)

$$P_e = \frac{1}{2} P(N(T) = 0) = \frac{1}{2} e^{-m_1} \qquad (16)$$

Letting $Pe = 10^{-9}$ gives $m_1 = 20.0$. The pulse must have an optical energy corresponding to an average of 20 photons to result in a bit error probability equal to 10^{-9}.

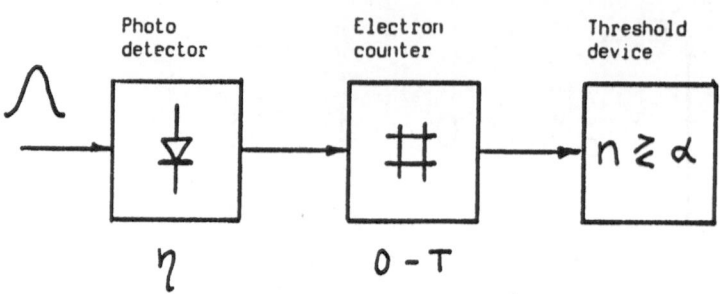

Figure 2:
Photon-counting receiver.

On the average half the time intervals contain optical pulses and the average number of photons per transmitted bit of information is

$$\frac{1}{2} m_1 = 10 \text{ photons/bit} \tag{17}$$

This quantity of 10 photons/bit is called the *quantum limit* for optical detection. It is a theoretical result assuming idealized conditions and represents a lower limit on the received optical power necessary in an optical communication system.

In the succeeding sections we will relax the idealized conditions assumed in the derivation of the quantum limit and study more realistic optical receivers.

5   Photon-counting receiver

In the derivation of the quantum limit both the signaling conditions and receiver operation are idealized. In this section we still assume the receiver to be ideal but we modify the received signal in order to be more realistic.

The receiver, shown in Fig. 2, consists of a photodetector with quantum efficiency η followed by a photoelectron counter and a threshold device.

The received optical signal is shown in Fig. 3. The pulses may now be broader than the bit-time interval T. This is typically caused by the dispersion resulting from transmission through an optical fiber. The overlapping of received pulses, which is called intersymbol interference (ISI), degrades receiver performance.

The received signal in Fig. 3 has a nonzero value P_0 during the "off" intervals which means that the light source is not switched off completely during the intervals representing binary zeroes. This is the case when a laser is operating

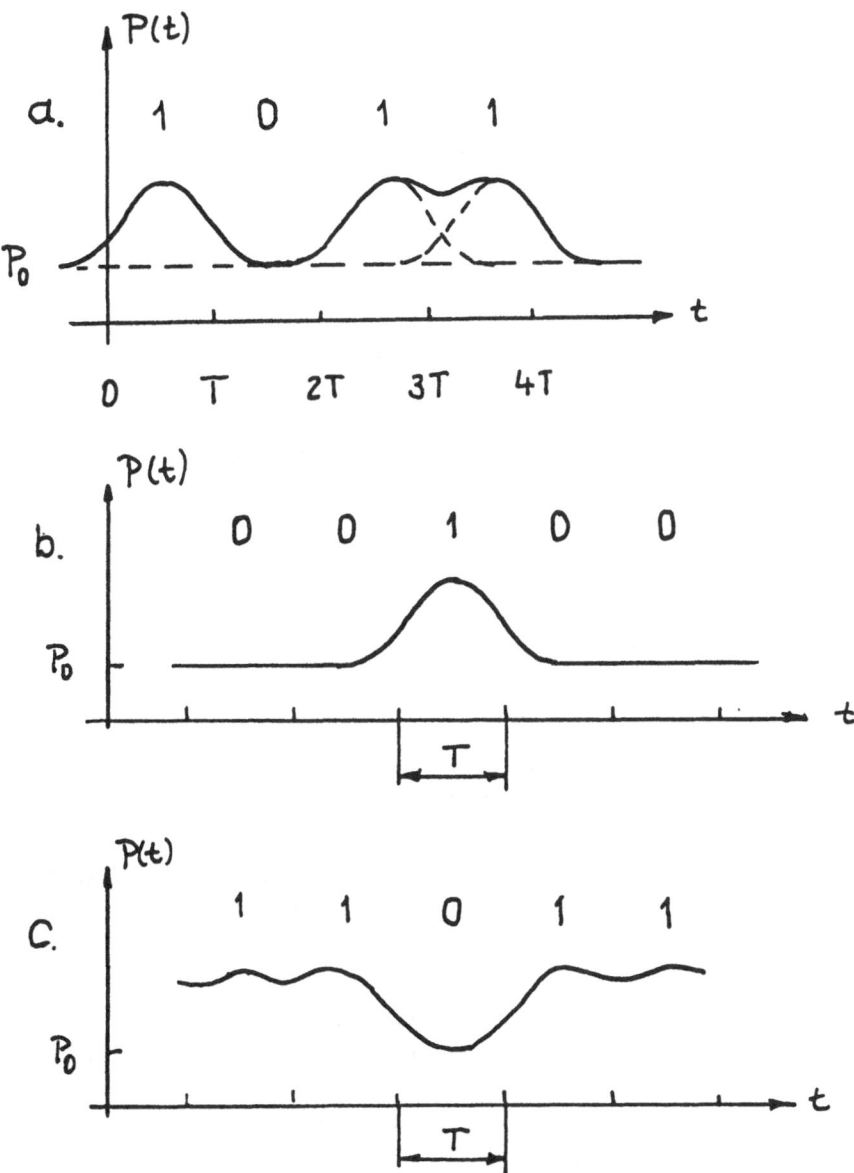

Figure 3:
Optical signals with intersymbol interference ISI.
a) General received signal.
b) Least favourable situation when a binary one is transmitted (minimal ISI).
c) Least favourable situation when a binary zero is transmitted (maximal ISI).

above its threshold. The same effect can also be caused by dark current in the photodetector.

We denote the received optical pulse representing a binary one by $p(t)$. The received optical signal $P(t)$ is

$$P(t) = P_0 + \sum_{k=-\infty}^{\infty} a_k p(t - kT) \tag{18}$$

where $a_k = 0$ or $a_k = 1$ depending on which binary symbol was transmitted.

The optical exposure of the detector by $P(t)$ generates a stream of photo-electrons with an intensity given by (3). In addition, the dark current i_0 of the detector also contributes to the number of electrons produced. The number of electrons occuring in the k:th signal interval has a Poisson distribution with mean value

$$m = \frac{i_0 T}{q} + \frac{\eta}{hf} \int_{kt}^{(k+1)T} P(t)dt \tag{19}$$

where i_0 represents the dark current of the photodetector.

From Fig. 3 it is evident that the part of $P(t)$ that falls into a certain symbol interval T depends on the symbol in that interval and also in general on the symbols in neighbouring intervals. This means that the detection probability will depend not only on the transmitted symbol but also on the symbols preceeding and succeeding it. We deal with this situation by considering the least favourable cases for detection.

When a binary one is transmitted in, say, time slot $k = 0$, the intersymbol interference can only increase the optical energy in the bit interval T. The smallest received optical energy and the largest error probability result when the symbol $a_0 = 1$ is surrounded by zeroes, i. e. $a_k = 0$, $k \neq 0$. See Fig. 3b. We design our receiver such that this error probability is less than a certain design value. The system will then operate at an error probability not greater than this value for all input data sequences.

It is convenient to normalize the optical pulse $p(t)$. Let its total optical energy be

$$b = \int_{-\infty}^{\infty} p(t)dt \tag{20}$$

and define the relative optical energy in the signaling interval $0 - T$

$$\delta = \frac{1}{b} \int_0^T p(t)dt \tag{21}$$

The mean corresponding to a transmitted binary one and minimal ISI is obtained from (19) and (18)

$$m_1 = \frac{i_0 T}{q} + \frac{\eta}{hf} (P_0 T + b\delta) \tag{22}$$

For detection of a binary zero the least favourable situation is when $a_0 = 0$ and it is surrounded by $a_k = 1$, $k \neq 0$ as illustrated in Fig. 3c which corresponds to maximal ISI.

The corresponding mean value is

$$
\begin{aligned}
m_0 & = \frac{i_0 T}{q} + \frac{\eta P_0 T}{hf} + \frac{\eta}{hf} \int_0^T \sum_{k \neq 0} p(t - kT) dt \quad (23) \\
& = \frac{i_0 T}{q} + \frac{\eta}{hf}[P_0 T + b(1 - \delta)]
\end{aligned}
$$

The receiver counts the number of photoelectrons during the symbol interval T. If the number observed is greater than a fixed threshold it decides that a binary one was transmitted and if it is lower or equal to the threshold a binary zero is announced. The receiver considered performs bit by bit processing. When ISI is present, a sequence-estimation receiver dealing with the entire received sequence (18) would produce a better result.

The number of photoelectrons is a Poisson distributed stochastic variable with a distribution (5). The receiver makes an erroneous decision when the observed number is greater than the threshold and a zero was transmitted and when it is lower or equal than the threshold and a binary one is transmitted. The error probability, with equal apriori probabilities for ones and zeros, is for all input sequences bounded by

$$
P_e \leq P(\alpha) = \frac{1}{2} \sum_{n=\alpha+1}^{\infty} \frac{m_0^n}{n!} e^{-m_0} + \frac{1}{2} \sum_{n=0}^{\alpha} \frac{m_1^n}{n!} e^{-m_1} \quad (24)
$$

where α denotes the value of the threshold.
The threshold α should be such that P_e is as low as possible. We choose α to be the value that minimizes $P(\alpha)$. Consider the difference

$$
\begin{aligned}
\Delta P(\alpha) & = P(\alpha) - P(\alpha - 1) \quad (25) \\
& = \frac{1}{2} \frac{m_1^\alpha}{\alpha!} e^{-m_1} - \frac{1}{2} \frac{m_0^\alpha}{\alpha!} e^{-m_0}
\end{aligned}
$$

Letting $\Delta P(\alpha) = 0$ and solving for α gives

$$
\alpha = \frac{m_1 - m_0}{\ln(m_1) - \ln(m_0)} \quad (26)
$$

It is easy to verify that the correct integer-valued threshold α is the integer part of (26)

$$
\alpha = INT\left(\frac{m_1 - m_0}{\ln(m_1) - \ln(m_0)}\right) \quad (27)
$$

Since m_0 and m_1 are the least favourable values that can occur for any sequence of input symbols, and since minimal and maximal ISI cannot occur simultaneously, the true error probability is certainly less than $P(\alpha)$ when intersymbol interference is present. For a system with no ISI the error probability $P_e = P(\alpha)$.

The error probability can be calculated from (24) using a table for the cumulative Poisson distribution, but this may be inconvenient in practice. In Appendix A upper bounds on the sums in (24) are studied. The following so-called improved Chernoff bounds are shown to constitute accurate approximations.

$$Pr(N \leq \alpha) = \sum_{n=0}^{\alpha} \frac{m}{n!} e^{-m} < \frac{1}{\sqrt{2\pi\alpha}} \frac{m}{m-\alpha} \exp[-\mu(m,\alpha)] \; ; \; m > \alpha \qquad (28)$$

$$Pr(N \geq \alpha) = \sum_{n=\alpha}^{\infty} \frac{m}{n!} e^{-m} < \frac{1}{\sqrt{2\pi\alpha}} \frac{\alpha+1}{\alpha-m+1} \exp[-\mu(m,\alpha)] \; ; \; m < \alpha \qquad (29)$$

where

$$\mu(m,\alpha) = m - \alpha[1 + \ln(m/\alpha)] \qquad (30)$$

The evaluation of error probability and the accuracy of (28) and (29) are illustrated by the following example.

Example 1 System without ISI

Consider a fiber optical system with an information rate of $R = 100 Mb/s$ working at $\lambda = 0.8\mu m$.
The received optical signal has two intensity levels $P_0 = 10^{-9}W$ for binary zero and $P_1 = 5 \cdot 10^{-9}W$ for binary one.
The photodetector has a quantum efficiency $\eta = 0.9$ and produces a dark current $i_0 = 1.5 \cdot 10^{-9}A$.
Calculate the bit error probability for a photon(photoelectron)-counting receiver.

Solution:

With optical power P_0 the average number of photoelectrons occuring in the time interval T is from (19)

$$m_p = \frac{\eta}{hf} \cdot P_0 \cdot T = \frac{0.9 \; 0.8 \cdot 10^{-6}}{6.626 \cdot 10^{-34} \cdot 2.998 \cdot 10^8} \cdot 10^{-9} \cdot 10^{-8} = 36.25$$

The dark current i_0 corresponds to an average number

$$m_i = \frac{i_0 T}{q} = \frac{1.5 \cdot 10^{-9} \cdot 10^{-8}}{1.602 \cdot 10^{-19}} = 93.63$$

of electrons during the symbol interval T.

The average number of electrons representing the binary zero symbol is

$$m_0 = m_p + m_i = 129.9$$

Since $P_1 = 5P_0$ the average number of electrons representing the binary one symbol is

$$m_1 = 5m_p + m_i = 274.9$$

The receiver threshold is calculated from (26) which gives $\alpha = 193.4$. The integer threshold for the electron counter is thus $\alpha = 193$.

The error probability calculated by the exact evaluation of the sums in formula (24) is

$$P_e = 1.0301 \cdot 10^{-7}$$

The discrete saddlepoint approximation presented in Appendix A gives

$$P_e = 1.0297 \cdot 10^{-7}$$

Replacing the summations in (24) by the upper bounds (28) and (29) result in

$$P_e < 1.054 \cdot 10^{-7}$$

which is a good approximation compared to the exact value and is easier to calculate.

□

A system with intersymbol interference is dealt with in the same way.

Example 2 System with ISI

Calculate the bit error probability for a system with the same received optical power as in Example 1 when the received pulses have Gaussian shape

$$p(t) = A \exp(-2t^2/T^2)$$

Solution:

The received signals now exhibit intersymbol interference (ISI). The received optical energy is

$$b = A \int_{-\infty}^{\infty} \exp(-2t^2/T^2)dt = AT\sqrt{\pi/2}$$

and the ISI parameter δ becomes (48)

$$
\begin{aligned}
\delta &= \frac{A}{b} \int_{-T/2}^{T/2} \exp(-2t^2/T^2)dt \\
&= \frac{1}{\sqrt{2\pi}} \int_{-1}^{1} \exp(-x^2/2)dx = 0.683
\end{aligned}
$$

The same received optical symbol energy as in Example 1 gives

$$b = (P_1 - P_0)T = 4 \cdot 10^{-9} \cdot 10^{-8} Ws$$

The mean value m_1 when symbol "one" is transmitted is from (22) for minimal ISI

$$
\begin{aligned}
m_1 &= \frac{i_0 T}{q} + \frac{\eta}{hf}(P_0 T + b\delta) \\
&= 93.63 + 3.625 \cdot 10^{18}(10^{-9} \cdot 10^{-8} + 4 \cdot 10^{-17} \cdot 0.683) = 228.9
\end{aligned}
$$

The mean value m_0 when symbol "zero" is transmitted is from (23) for maximal ISI

$$
\begin{aligned}
m_0 &= \frac{i_0 T}{q} + \frac{\eta}{hf}[P_0 T + b(1 - \delta)] \\
&= 93.63 + 3.625 \cdot 10^{18}[(10^{-9} \cdot 10^{-8} + 4 \cdot 10^{-17}(1 - 0.683)] = 175.8
\end{aligned}
$$

The integer threshold corresponding to those values of m_1 and m_0 is from (27)

$$\alpha = 201$$

which results in the bit error probability

$$
\begin{aligned}
P_e < P(\alpha) &= 3.07 \cdot 10^{-2} \text{ exact} \\
P_e < P(\alpha) &\approx 3.03 \cdot 10^{-2} \text{ saddlepoint approximation} \\
P_e < P(\alpha) &< 3.59 \cdot 10^{-2} \text{ approx. using (28) and (29)}
\end{aligned}
$$

□

A comparison with Example 1 shows a considerable degradation in performance from the intersymbol interference.

For increasing m the Poisson distribution asymptotically approaches a Gaussian distribution. As discussed in Appendix A the approximation is not very accurate in the tails of the distribution. Nevertheless, the Gaussian density is often used to obtain an approximate expression for the error probability for an ideal photon-counting receiver.

A Poisson distribution with mean m has variance $\sigma^2 = m$, see (6). Approximating the sums in (24) by integrals of Gaussian density functions gives

$$P_e \approx \frac{1}{2} Q\left(\frac{m_1 - \alpha}{\sqrt{m_1}}\right) + \frac{1}{2} Q\left(\frac{\alpha - m_0}{\sqrt{m_0}}\right) \tag{31}$$

with

$$Q(x) = \frac{1}{\sqrt{2\pi}} \int_x^\infty e^{-s^2/2} ds \tag{32}$$

To obtain a simple expression let the threshold α be determined by making the two terms in (31) equal in magnitude, i.e.

$$\frac{m_1 - \alpha}{\sqrt{m_1}} = \frac{\alpha - m_0}{\sqrt{m_0}}$$

which result in

$$\alpha = \sqrt{m_1 m_0} \tag{33}$$

Substitution of (33) into (31) gives

$$P_e \approx Q(\sqrt{m_1} - \sqrt{m_0}) \tag{34}$$

We illustrate the approximation by applying it to the previous two examples

Example 3 Gaussian Approximation

Calculate the error probabilities in Examples 1 and 2 using (34).

Solution:

The Q-function (32) can be evaluated from a table over the Gaussian distribution functions. A simple approximate expression is the improved Chernoff bound derived in Appendix B.

$$Q(x) < \frac{1}{\sqrt{2\pi}x} \exp(-x^2/2) \tag{35}$$

a) $m_0 = 129.9$ and $m_1 = 274.9$ gives

$$x = \sqrt{274.9} - \sqrt{129.9} = 5.18$$

The approximation (35) yields

$$P_e \approx Q(x) < 1.13 \cdot 10^{-7}$$

b) $m_0 = 175.8$ and $m_0 = 228.9$ gives

$$x = \sqrt{228.9} - \sqrt{175.8} = 1.87$$

From (35)

$$P_e \approx Q(x) < 3.71 \cdot 10^{-2}$$

□

The Gaussian approximation works well. This is partly due to the fact that it overestimates the probability of the lower tail and underestimates the upper tail of the Poisson distribution and these two errors tend to cancel each other. See Appendix A.

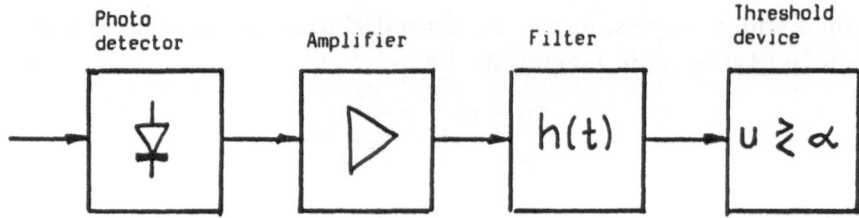

Figure 4:

An optical receiver containing a linear filter followed by a threshold device.

6 Receiver with Rectangular Filter and Additive Thermal Noise

The receivers studied in sections 4 and 5 are ideal in the sense that they are assumed to be capable of detecting and counting single photons. This requires a perfect detector without internal noise or any other deficiencies. A practical receiver usually consists of a photodetector followed by a low noise amplifier and a suitable filter before a threshold device. See Fig. 4.

In this section we consider a receiver with a special kind of filter, the so-called integrate-and-dump filter. Such a filter [11] consists in principle of an integrator which is reset after T seconds.

The integrate-and-dump filter has an impulse response of rectangular shape as depicted in Fig. 5. The choice of an integrate-and-dump filter is mainly for simplifying the analysis. It illustrates the properties of a realistic receiver without having to devote too much effort to mathematical details.

We begin the analysis of the integrate-and-dump filter receiver assuming that the amplifier and the filter are noiseless. The photodetector sends a stream of photoelectrons to the amplifier. In the shot noise model (7) each electron generates a current pulse of shape $g(t)$, which we assume to have a duration much shorter than the symbol interval duration T. The amplified current $i(t)$ at the filter input is from (7) and (8)

$$i(t) = A \sum_{k=-\infty}^{\infty} g(t - t_k) \tag{36}$$

with t_k generated by a Poisson process.

The output of the rectangular filter at time $t = T$ is

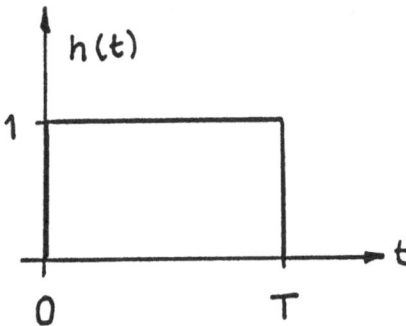

Figure 5:

The impulse response of an integrate-and-dump (rectangular) filter.

$$u(t = T) \;=\; \int_{-\infty}^{\infty} h(T-t)i(t)dt = \int_0^T i(t)dt \qquad (37)$$
$$\approx \; N \cdot A \cdot q$$

where N is the number of photoelectrons, i.e. the number of Poisson events in the current $i(t)$ during the time interval $[0,\,T]$, and q is the electron charge. An ideal integrate-and-dump filter receiver without thermal noise is thus equivalent to the photon-counting receiver analysed in section 5.

To study the effect of thermal noise in the receiver we introduce an additive noise component at the input of the filter

$$i(t) = A \sum g(t - t_k) + i_n(t) \qquad (38)$$

The current $i_n(t)$ represents the thermal noise from the photodetector, the amplifier and the filter. It is assumed to be a zero-mean, white and Gaussian stochastic process.

Integration of (38) gives

$$u(t = T) \;=\; \int i(t)dt \qquad (39)$$
$$\approx \; NAq + \xi$$

where ξ is a zero mean Gaussian stochastic variable. Its variance is

$$\sigma_\xi^2 = E\left\{ \left(\int_0^T i_n(t)dt \right)^2 \right\} = R_0 T \qquad (40)$$

with R_0 equal to the (two-sided) spectral density of $i_n(t)$.

The amplification factor A can be changed if the threshold is adjusted accordingly. For convenience we let $Aq = 1$ and normalize $i_n(t)$ in such a way that the decision variable (39) becomes

$$U = N + X \tag{41}$$

The stochastic variable U is the sum of two independent variables. The component N has a Poisson distribution (5) and X is Gaussian. The statistics of U are studied in Appendix B where efficient methods of evaluating the probability of U falling above or or below a threshold are presented. The saddlepoint approximation presented in Appendix C is an accurate and convenient method for estimation of error probabilites. We illustrate the calculations with an example.

Example 4 Saddlepoint approximation

Consider the optical system described in Example 1 with $m_0 = 129.9$ and $m_1 = 274.9$.

Assume that Gaussian additive noise corresponding to normalized value of $\sigma = 5$ is present in the detector.
Determine the detector threshold α and calculate the error probability.

Solution :

The error probability is

$$P_e = \frac{1}{2}[P(U > \alpha/m = m_0) + P(U < \alpha/m = m_1)] \tag{42}$$

Using the saddlepoint approximations (B.17) and (B.21) for the probabilities in (42) yields

$$P_e \approx \frac{1}{2}[q_+(\alpha, \beta_0) + q_-(\alpha, \beta_1)] \tag{43}$$

where from (C.8) and (C.11)

$$q_+(\alpha, \beta_0) = \frac{\exp[m_0(e^{\beta_0} - 1) + \beta_0^2\sigma^2/2 - \beta_0\alpha]}{\sqrt{2\pi(m_0 e^{\beta_0} + \sigma^2 + 1/\beta_0^2)}} \tag{44}$$

and

$$q_-(\alpha, \beta_1) = \frac{\exp[m_1(e^{\beta_1} - 1) + \beta_1^2\sigma^2/2 - \beta_1\alpha]}{\sqrt{2\pi(m_1 e^{\beta_1} + \sigma^2 + 1/\beta_1^2)}} \tag{45}$$

The parameters β_0 and β_1 are determined by (B.19)

$$m_0 e^{\beta_0} + \sigma^2\beta_0 - \alpha - 1/\beta_0 = 0 \tag{46}$$

and

$$m_1 e^{\beta_1} + \sigma^2\beta_1 - \alpha - 1/\beta_1 = 0 \tag{47}$$

respectively.

The threshold α is determined such that (43) is minimal. This must be done numerically. For $m_0 = 129.9$, $m_1 = 274.9$ and $\sigma = 5$ the best threshold turns out to be

$$\alpha = 194.04$$

resulting in an error probability (43) of

$$P_e \approx 5.153 \cdot 10^{-7}$$

An exact evaluation of (42) gives the best threshold

$$\alpha = 194.05$$

and an error probability of

$$P_e = 5.156 \cdot 10^{-7}$$

□

The example shows that the saddlepoint approximation gives extremely accurate numerical results for an optical receiver with both Poisson and Gaussian noise.

The calculations above are simple to carry out on a digital computer but the numerical optimizations required to determine α, β_0 and β_1 may be cumbersome to perform on a hand calculator.
A fairly accurate estimate of the error probability is obtained if the distribution for the compound variable U is approximated by a Poisson or a Gaussian distribution. We will study both these probabilities.

Thermal noise is the result of Brownian motions of the electrons. It can be modeled as a shot noise process of the same nature as the photon process described in section 3.1. This motivates the approximation of the thermal noise variable X as an independent Poisson process with variance equal to σ^2.

The sum of two independent Poisson variables has a Poisson distribution and the distribution for $U = N + X$ is represented by a single Poisson distribution with mean $m + \sigma^2$.
The receiver threshold can now be determined from (27) and the error probability by (28) and (29).

Example 5   Poisson approximation

The same assumptions as in Example 4 give $m_0 = 129.9 + 25$ and $m_1 = 274.9 + 25$. The corresponding threshold is

$$\alpha_p = \frac{299.9 - 154.9}{\ln(299.9) - \ln(254.9)} = 219.5 \tag{48}$$

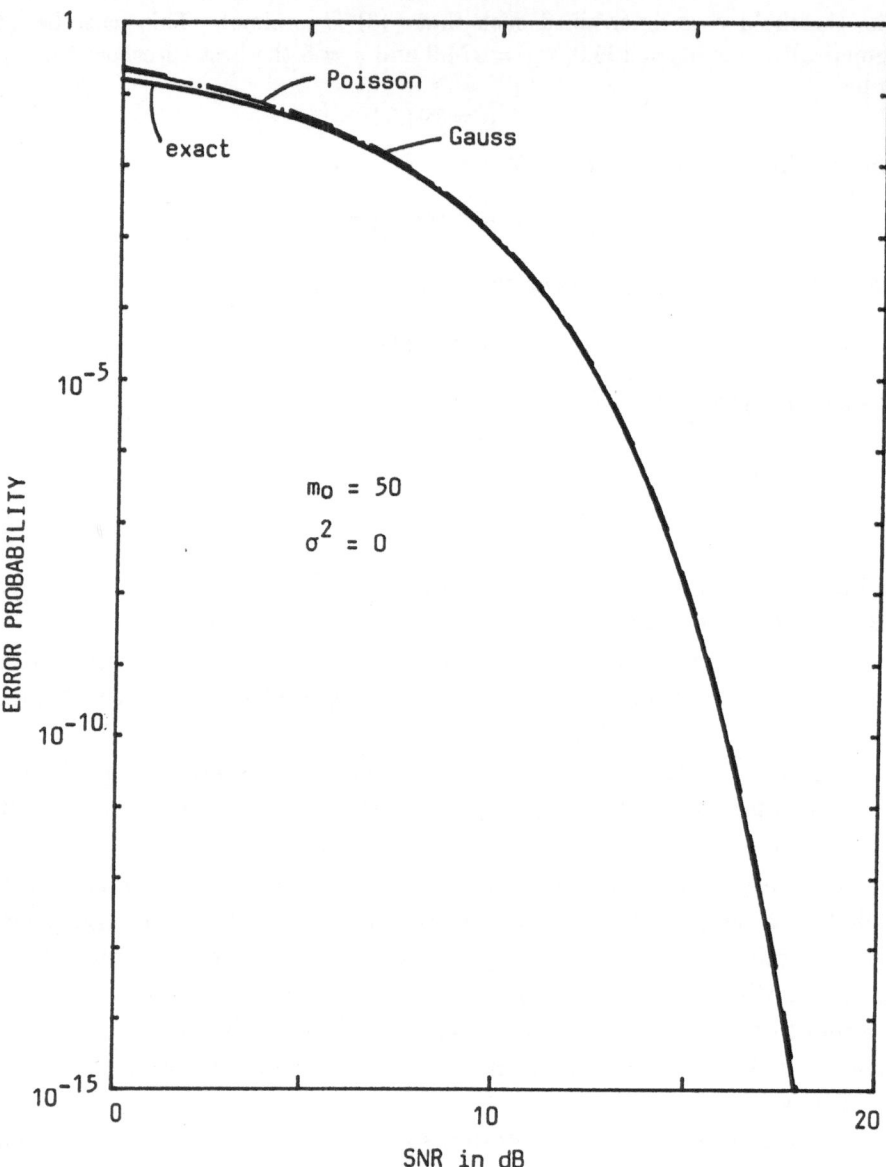

Figure 6:
Transmission error probability for an optical system with rectangular filter receiver. The exact value is shown together with the Gaussian and Poisson approximations.

a) Poisson channel (no thermal noise).

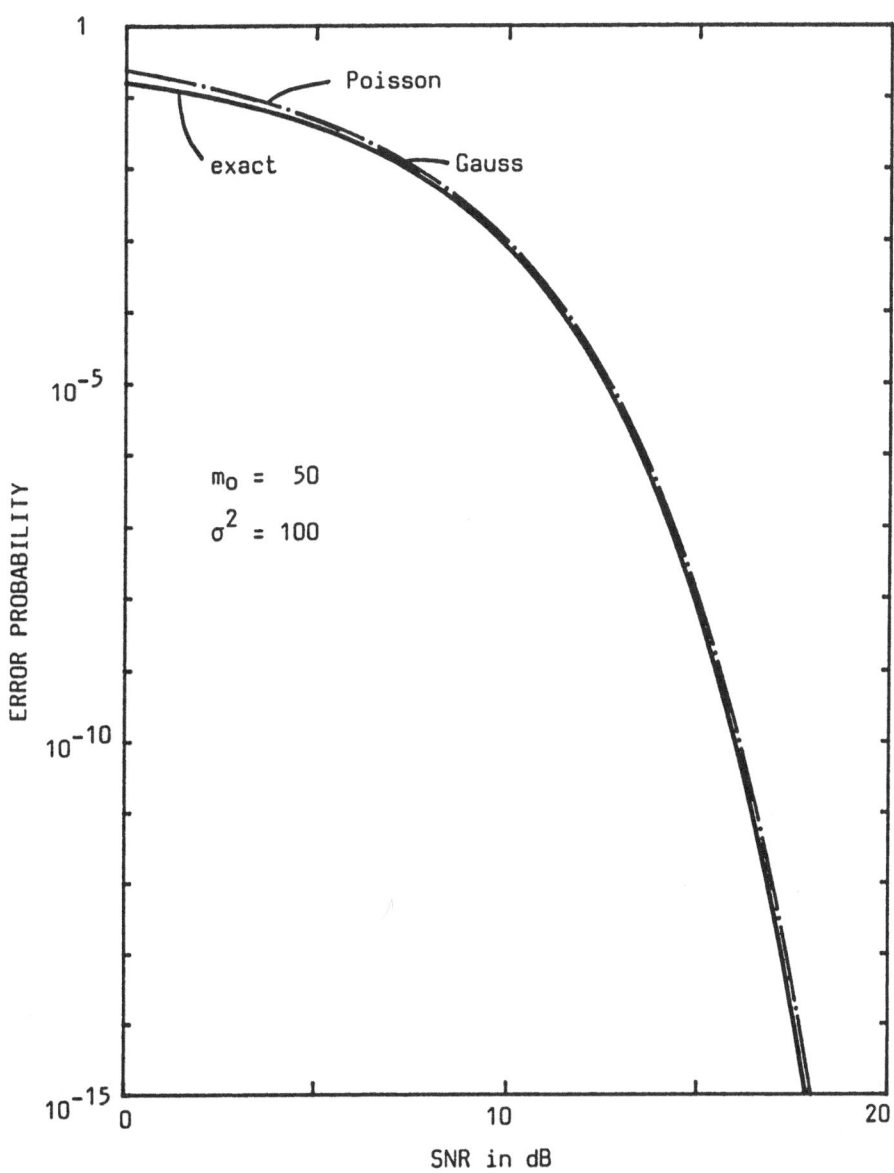

Figure 6:

b) Poisson plus Gaussian noise.

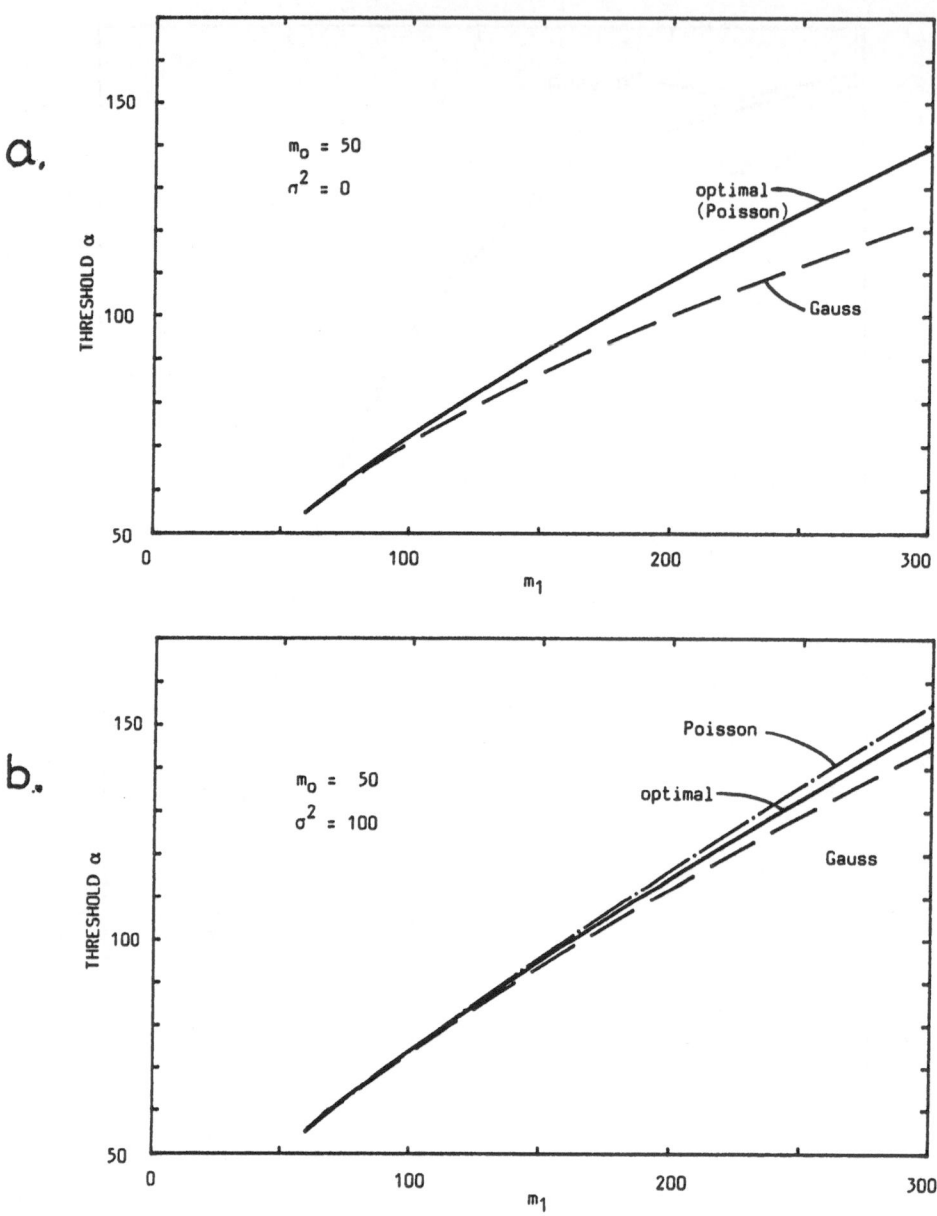

Figure 7:

Detector threshold for a receiver with rectangular filter.
The optimal value is shown together with the thresholds
produced by the Gaussian and the Poisson approximations.

a) Poisson channel (no thermal noise).
b) Poisson plus Gaussian noise.

To compare this with the correct threshold $\alpha = 194.05$ from Example 4 the mean value σ^2 of the thermal noise Poisson process has to be subtracted which gives

$$\alpha = 219.47 - 25 = 194.5$$

The error probability calculated from the approximate expressions (28) and (29) with the threshold (48) is

$$P_e \approx 6.47 \cdot 10^{-7}$$

The correct threshold $\alpha = 194.05$ gives

$$P_e \approx 6.53 \cdot 10^{-7}$$

□

The other alternative, to approximate the Poisson variable N by a Gaussian distribution, is common in the literature. The compound variable is then considered to be a Gaussian variable with mean m and variance $m + \sigma^2$.
The error probability is, compare with (31)

$$P_e \approx \frac{1}{2}\left[Q\left(\frac{m_1 - \alpha}{\sqrt{m_1 + \sigma^2}}\right) + Q\left(\frac{\alpha - m_0}{\sqrt{m_0 + \sigma^2}}\right)\right] \tag{49}$$

with the Q-function defined in (32).

The threshold determined by making the two terms in (49) equal in magnitude is

$$\alpha_g = \frac{m_1\sqrt{m_0 + \sigma^2} + m_0\sqrt{m_1 + \sigma^2}}{\sqrt{m_1 + \sigma^2} + \sqrt{m_0 + \sigma^2}} \tag{50}$$

Substitution into (49) gives

$$Pe \approx Q\left(\frac{m_1 - m_0}{\sqrt{m_1 + \sigma^2} + \sqrt{m_0 + \sigma^2}}\right) \tag{51}$$

which constitutes a direct relation between the system parameters m_0, m_1 and σ and the transmission error probability

Example 6   Gaussian approximation

The values $m_0 = 129.9$, $m_1 = 274.9$ and $\sigma^2 = 25$ gives

$$\alpha_g = 190.5$$

and the error probability (51) is

$$P_e \approx 5.75 \cdot 10^{-7}$$

when the simple approximate expression (35) is used for the Q-function.

The threshold obtained from (50) is not the correct one. If the correct value $\alpha = 194.05$ is used together with the Gaussian approximation the error probability is

$$P_e \approx 8.24 \cdot 10^{-7}$$

□

Both the Poisson and the Gaussian approximations work well. Fig 6.a shows that the Gaussian approximation is an acceptable estimate of the system error probability even for $\sigma^2 = 0$, i. e. with no thermal noise present. The Poisson approximation is of course more exact in this case. A typical case with both optical and thermal noise is presented in Fig 6.b which shows almost identical results for both approximations. The approximate threshold (26) makes the exponents in (28) and (29) equal and the numerical complexity is about the same for both.

For small values of σ the Poisson approximation produces a receiver threshold closer to the correct one than the Gaussian approximation as is illustrated by Figs 7.a and 7.b.

APPENDIX A

BOUNDS AND APPROXIMATIONS FOR THE POISSON DISTRIBUTION

A.1   Chernoff bounds

The Chernoff bound [3] is a simple and quite general bound on the tail of a probability distribution

$$P(X \geq \alpha) = \int_{\alpha}^{\infty} p(x)dx \tag{A.1}$$

The integral can be written as

$$\int_{-\infty}^{\infty} u(x - \alpha)p(x)dx \tag{A.2}$$

where $u(x)$ is the unit step function.

Since $u(x) \leq \exp(sx)$ for $s \geq 0$ is

$$P(X \geq \alpha) \leq \int_{-\infty}^{\infty} \exp[s(x - \alpha)]p(x)dx$$

$$= e^{-s\alpha} \int_{-\infty}^{\infty} e^{sx}p(x)dx \tag{A.3}$$

The last integral in (A.3) is equal to the moment-generating function

$$\Psi(s) = E\{\exp(sX)\} \tag{A.4}$$

and the bound can be written as

$$P(X \geq \alpha) \leq e^{-s\alpha}\Psi(s) \, ; \quad s \geq 0 \tag{A.5}$$

The value of s which minimizes the right-hand side of (A.5) results in the tightest bound and is obtained by setting the derivative of (A.5), with respect to s, equal to zero [3]. This gives the equation

$$\alpha\Psi(s) = \Psi'(s) \tag{A.6}$$

The lower tail is bounded in the same way, since for $s \leq 0$

$$P(X \leq \alpha) = \int_{-\infty}^{\alpha} p(x)dx$$

$$= \int_{-\infty}^{\infty} u(\alpha - x)p(x)dx$$

$$\leq \int_{-\infty}^{\infty} \exp[s(x - \alpha)]p(x)dx$$

$$= e^{-s\alpha}\Psi(s) \, ; \quad s \leq 0 \tag{A.7}$$

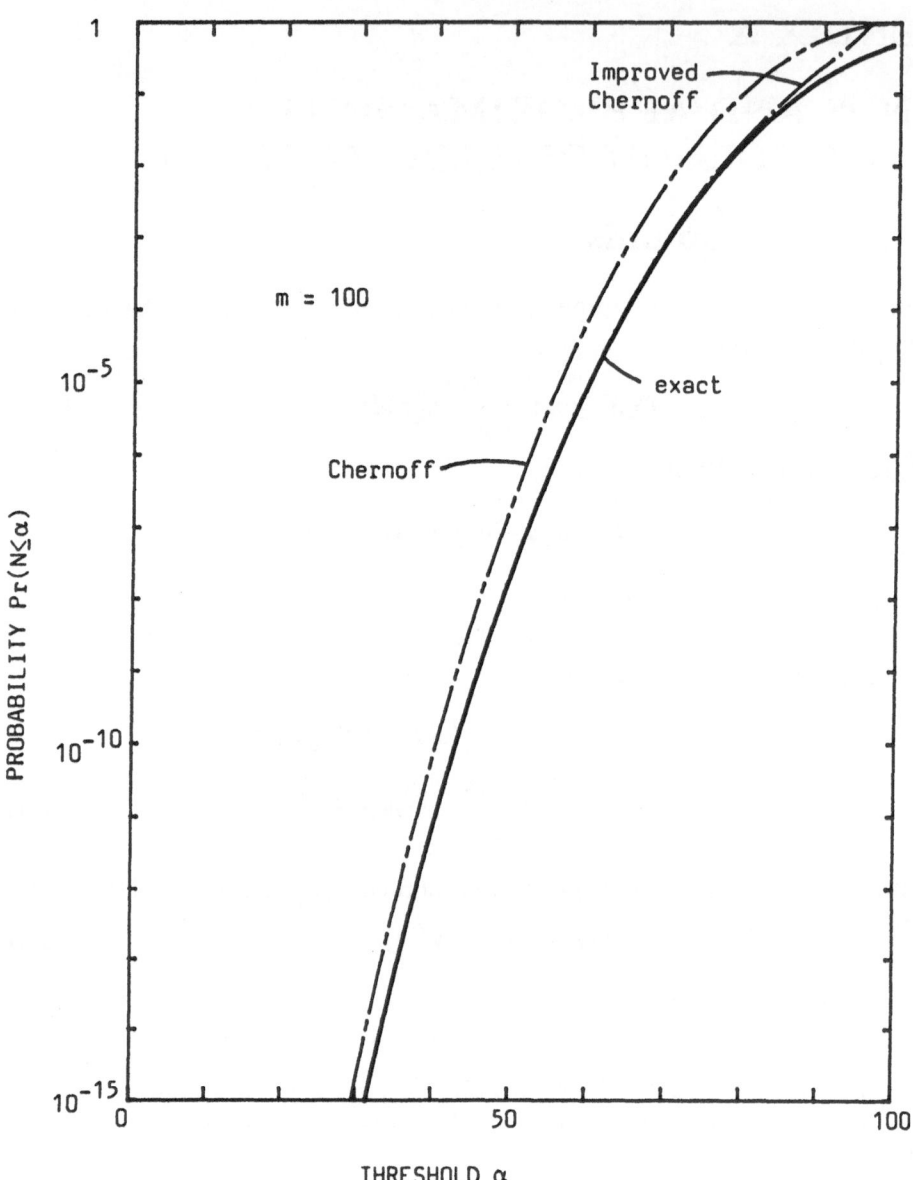

Figure 8:
Cumulative probabilities for a Poisson variable.
Exact value and upper bounds.
a) lower tail.

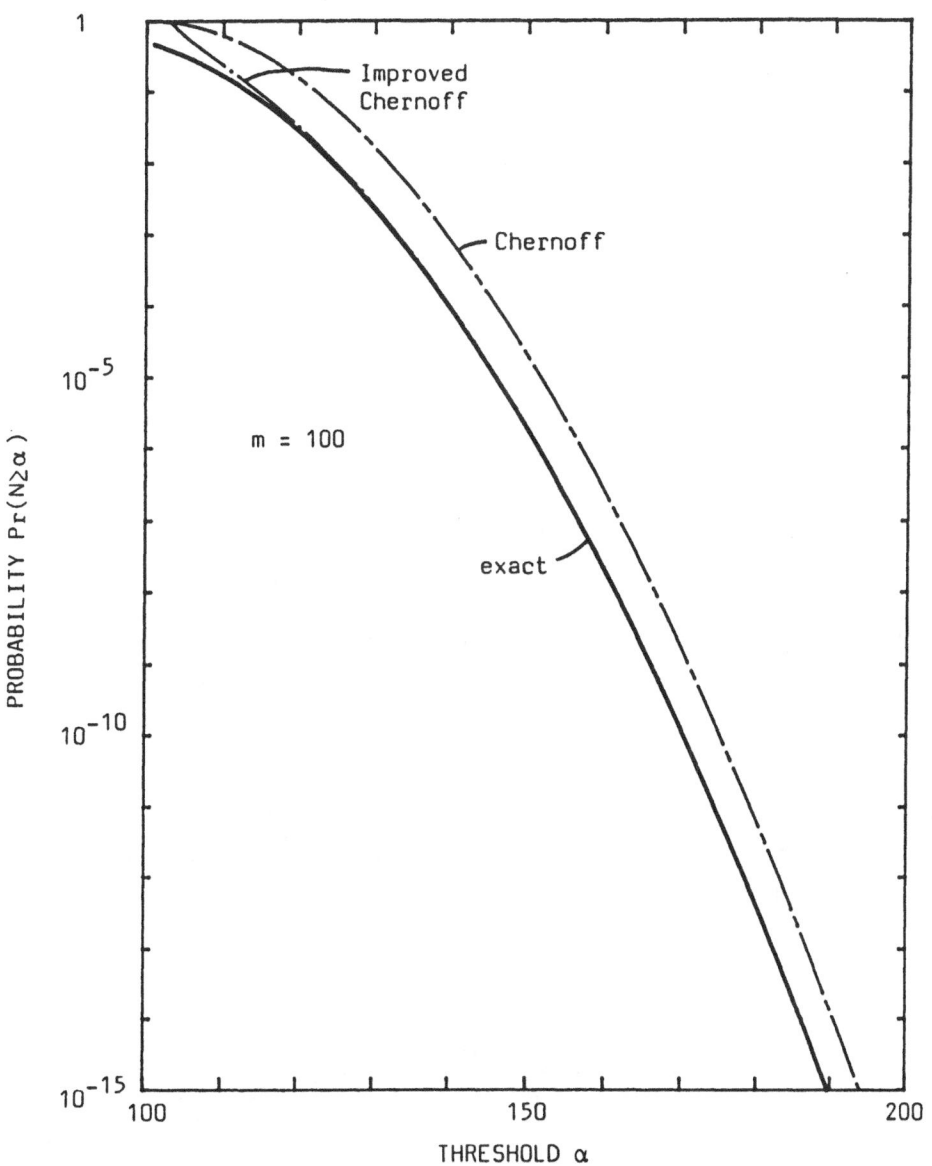

Figure 8:

b) upper tail.

For the Poisson distribution

$$\Psi(s) = E\{e^{sN}\} = \sum_{n=0}^{\infty} \frac{m^n e^{-m}}{n!} e^{sn}$$

$$= e^{-m} \sum_{n=0}^{\infty} \frac{(me^s)^n}{n!}$$

$$= \exp[m(e^s - 1)] \tag{A.8}$$

Substitution into (A.6) gives

$$\alpha = me^s \tag{A.9}$$

Substitution of (A.8) and (A.9) into (A.5) and (A.7) gives the Chernoff bounds for the Poisson distribution

$$P(N \geq \alpha) \leq (m/\alpha)^{\alpha} \exp(\alpha - m); \quad \alpha \geq m \tag{A.10}$$
$$P(N \leq \alpha) \leq (m/\alpha)^{\alpha} \exp(\alpha - m); \quad \alpha \leq m \tag{A.11}$$

Fig 8 gives examples of how these bounds compare with the exact probabilities.

A.2   Improved Chernoff bounds

A Chernoff bound is exponentially tight [11] which means that it is an asymptotically correct expression for $\log(P)$. It is evident from Fig 8 that it nevertheless can give a substantial error when used to estimate the error probability.

An improved bound for the Poisson distribution is obtained as follows [7]

$$\sum_{n=\alpha}^{\infty} e^{-m} \frac{m^n}{n!} = e^{-m} \frac{m^{\alpha}}{\alpha!} \left(1 + \frac{m}{\alpha+1} + \frac{m^2}{(\alpha+1)(\alpha+2)} + \cdots \right)$$

$$< e^{-m} \frac{m^{\alpha}}{\alpha!} \left(1 + \frac{m}{\alpha+1} + (\frac{m}{\alpha+1})^2 + \cdots \right)$$

$$= e^{-m} \frac{m^{\alpha}}{\alpha!} \frac{\alpha+1}{\alpha-m+1}; \quad \alpha > m \tag{A.12}$$

The Stirling approximation [1]

$$\alpha! > \sqrt{2\pi\alpha} \, \alpha^{\alpha} e^{-\alpha} \tag{A.13}$$

gives

$$P(N \geq \alpha) < \frac{1}{\sqrt{2\pi\alpha}} \frac{\alpha+1}{\alpha-m+1} (m/\alpha)^{\alpha} \exp(\alpha - m); \quad \alpha > m \tag{A.14}$$

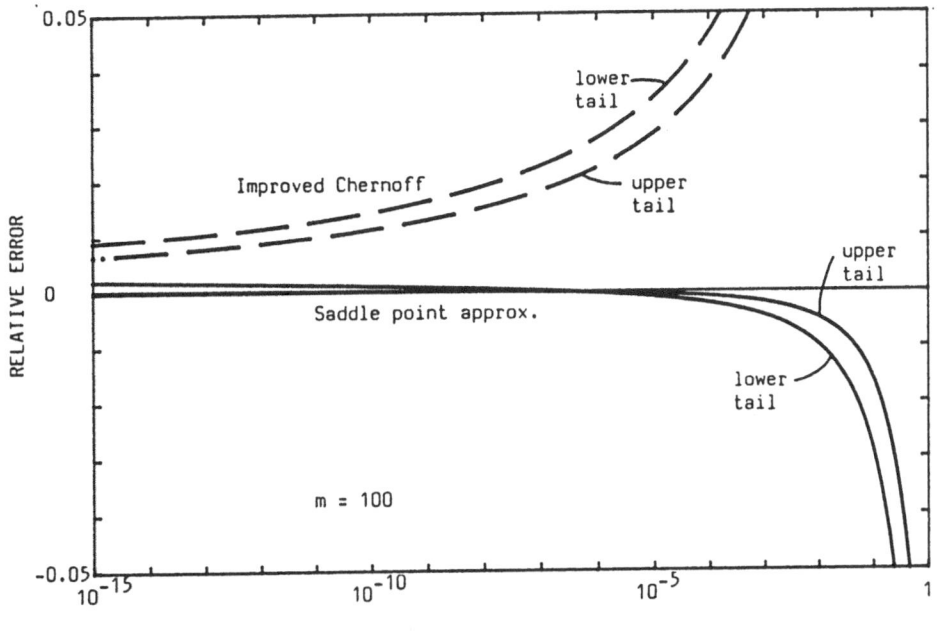

Figure 9:
Accuracy of the discrete saddlepoint approximation and of the improved Chernoff bounds for the cumulative Poisson distribution.
The diagram shows the relative error $(P - P_{exact})/P_{exact}$.

In the same way

$$
\begin{aligned}
\sum_{n=0}^{\alpha} e^{-m} \frac{m^n}{n!} &= e^{-m} \frac{m^\alpha}{\alpha!} \left(1 + \frac{\alpha}{m} + \frac{\alpha(\alpha-1)}{m^2} + \cdots + \frac{\alpha!}{m^\alpha} \right) \\
&< e^{-m} \frac{m^\alpha}{\alpha!} \left(1 + \frac{\alpha}{m} + (\frac{\alpha}{m})^2 + \cdots + (\frac{\alpha}{m})^\alpha \right) \\
&= e^{-m} \frac{m^\alpha}{\alpha!} \frac{1 - (\alpha/m)^{\alpha+1}}{1 - \alpha/m} \\
&< e^{-m} \frac{m^\alpha}{\alpha!} \frac{m}{m - \alpha} \,; \quad \alpha < m
\end{aligned}
\tag{A.15}
$$

The Stirling approximation (A.13) gives

$$
P(N \leq \alpha) < \frac{1}{\sqrt{2\pi\alpha}} \frac{m}{m - \alpha} (m/\alpha)^\alpha \exp(\alpha - m) \,; \quad \alpha < m
\tag{A.16}
$$

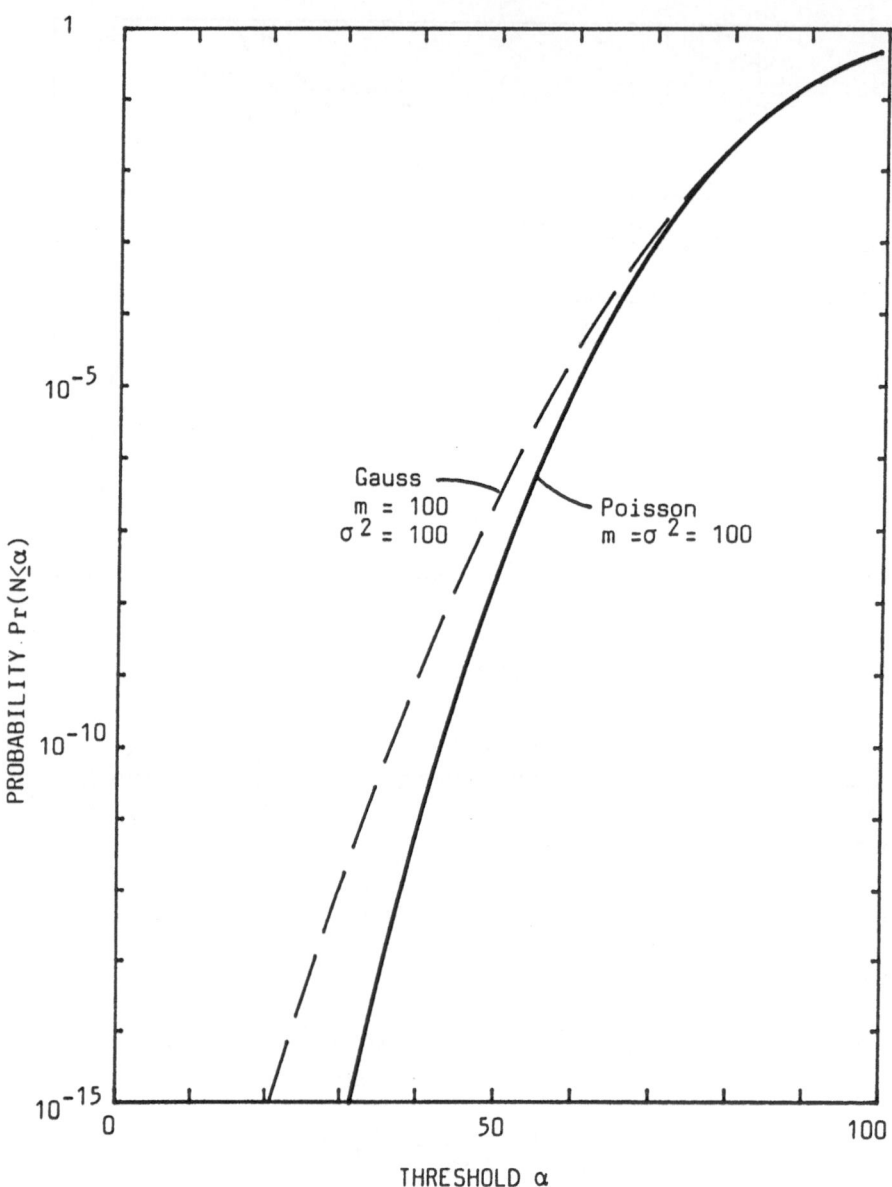

Figure 10:

Poisson and Gaussian distributions with the same
mean $m = 100$ and variance $\sigma^2 = 100$.

a) lower tail.

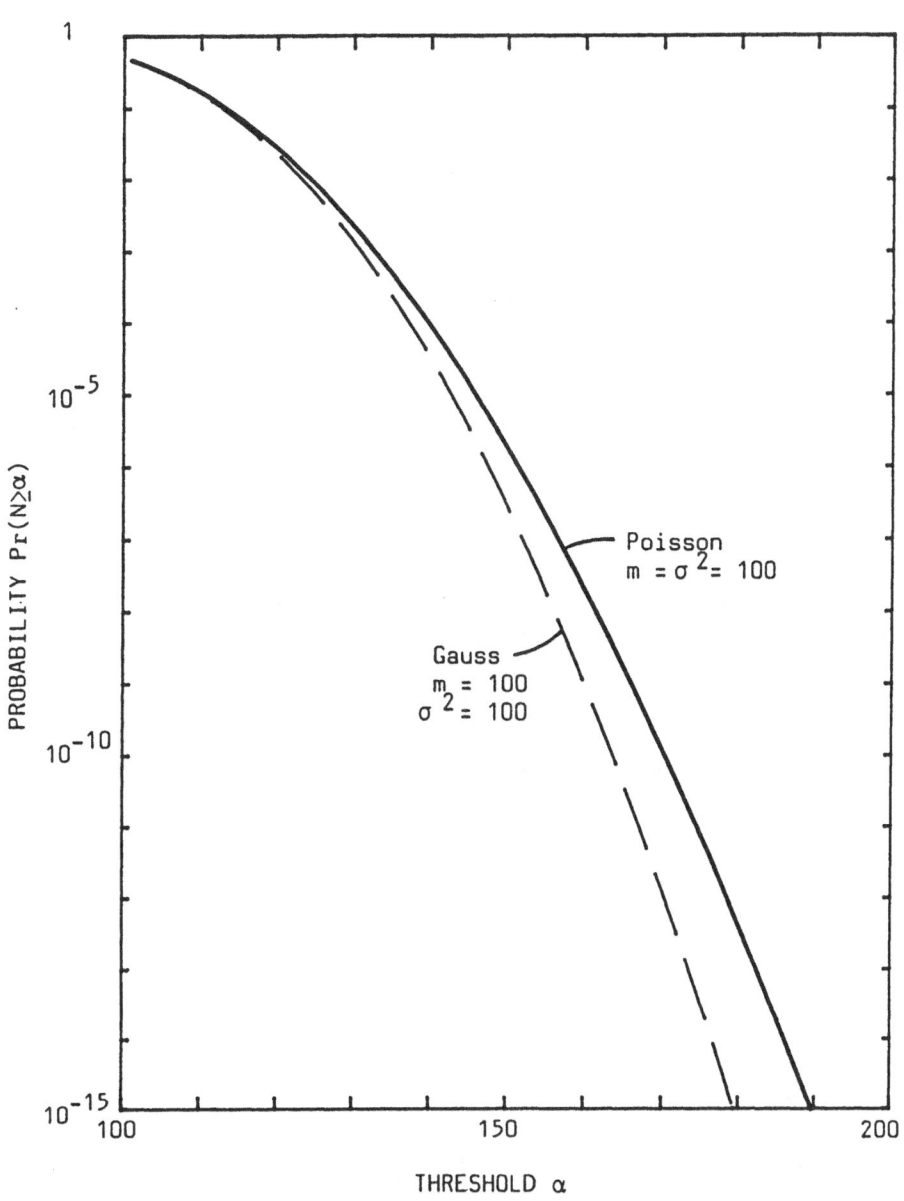

Figure 10:

b) upper tail.

The improved bounds (A.14) and (A.16) are illustrated in Fig 8. They have the same exponent as the original bounds (A.10) and (A.11) but they are much closer to the exact values. The relative error for the bounds is shown in Fig 9 to be only a few percent in the region of interest for detection probabilities.

A.3   Saddlepoint approximation

A convenient and efficient way of estimating the tails of a probability distribution is the saddlepoint method described in Appendix C.

The upper tail of a discrete probability distribution is approximately equal to (C.18)

$$P(N \geq \alpha) \approx [2\pi \phi''(\beta_0)]^{-1/2} \exp[\phi(\beta_0)] \tag{A.17}$$

with, see (C.17),

$$\phi(z) = \ln[\Phi(z)] - \alpha \ln(z) - \ln(|z - 1|) \tag{A.18}$$

where $\Phi(z)$ is the generating function (C.12) of the distribution.

For the Poisson distribution is, c. f. (A.8)

$$\Phi(z) = E\{z^N\} = \exp[m(z - 1)] \tag{A.19}$$

Substitution of (A.19) into (A.18) gives

$$\phi(z) = m(z - 1) - \alpha \ln(z) - \ln(|z - 1|) \tag{A.20}$$

The first derivative of $\psi(z)$ is

$$\phi'(z) = m - \alpha/z - 1/(z - 1) \tag{A.21}$$

and the second-order derivative

$$\phi''(z) = \alpha/z^2 + 1/(z - 1)^2 \tag{A.22}$$

The parameter β_0 is obtained by solving the equation $\phi'(z) = 0$. It is a second-order algebraic equation with the solution

$$\beta_0 = \frac{\alpha + m + 1}{2m} \pm \sqrt{\left(\frac{\alpha + m + 1}{2m}\right)^2 - \frac{\alpha}{m}} \tag{A.23}$$

For the upper tail is $\beta_0 > 1$ which corresponds to the positive sign in (A.23).

For the lower tail the saddlepoint approximation is

$$P(N \leq \alpha) \approx [2\pi \phi''(\beta_0)]^{-1/2} \exp[\phi(\beta_0)] \tag{A.24}$$

with $\beta_0 < 1$ which means that the minus sign in (A.23) now applies.

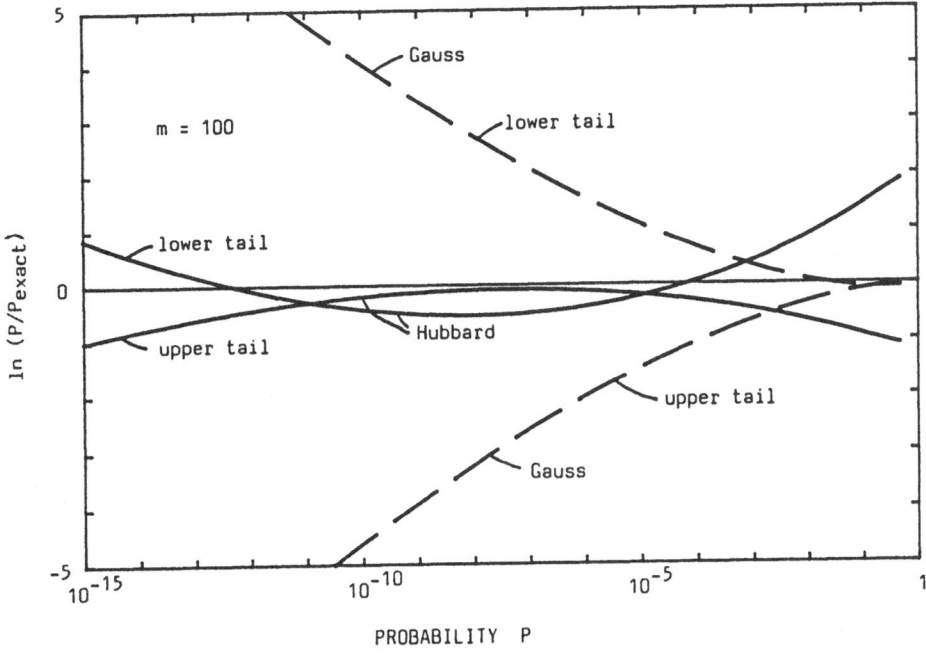

Figure 11:
Accuracy of the Gaussian distribution and of the modified Gaussian distribution of Hubbard [5] approximating a Poisson distribution.
The diagram shows the logarithmic deviation in cumulative probability.

The saddlepoint approximation of the Poisson distribution gives accurate estimates. It is illustrated in Fig 9 which shows a relative error of 0.2 percent or less for probabilities between 10^{-4} and 10^{-15}.

A.4   Gaussian approximation

In textbooks and papers the error probability of optical communication systems is often calculated by approximating the Poisson distribution with a Gaussian distribution. Fig 10 illustrates this approach showing the upper and lower tails of Poisson and Gaussian distributions having the same mean and variance.

The Poisson distribution becomes more "Gaussian" with increasing mean but the probability mass in the tails is not accurately estimated by the Gaussian cumulative distribution. A better estimate is obtained by introducing a nonlinear relation between the parameters of the Poisson and the Gaussian distributions. Many such approximations can be found in the literature. As an example Fig 11 shows an approximation [5] suggested for evaluation of optical communication systems [6]. The diagram shows $\ln(P/P_{exact})$, which for arguments close to unity is equal to the relative error.

The approximation is accurate to within a factor of two for probabilities between 10^{-3} and 10^{-13} which is a much larger relative error than for the saddlepoint approximation or the improved Chernoff bound.

APPENDIX B

BOUNDS AND APPROXIMATIONS FOR THE POISSON-PLUS-GAUSS DISTRIBUTION

Consider the sum of two independent stochastic variables

$$U = N + X \tag{B.1}$$

The component N has a Poisson distribution (5) with mean m and X is Gaussian with zero mean and variance σ^2. The probability density of U is the convolution of the densities of N and X

$$p(u) = \sum_{n=0}^{\infty} \frac{e^{-m}}{\sqrt{2\pi\sigma^2}} \frac{m^n}{n!} \exp[-(u-n)^2/2\sigma^2)] \tag{B.2}$$

The infinite summation makes (B.2) difficult to handle analytically and numerically. A more tractable way is to utilize the moment-generating function which, since U is the sum of two independent variables, is the product of the generating functions of the components

$$\Psi_u(s) = \Psi_n(s) \cdot \Psi_x(s) \tag{B.3}$$

The moment-generating function for a Poisson variable with mean m is from (A.8)

$$\Psi_n(s) = \exp[m(e^s - 1)] \tag{B.4}$$

For a Gaussian variable with zero mean and variance σ^2 is

$$\Psi_x(s) = \frac{1}{\sqrt{2\pi\sigma^2}} \int_{-\infty}^{\infty} e^{sx} e^{-x^2/s\sigma^2} dx = \exp(s^2\sigma^2/2) \tag{B.5}$$

Substitution of (B.4) and (B.5) into (B.3) gives

$$\Psi_u(s) = \exp[m(e^s - 1) + s^2\sigma^2/2] \tag{B.6}$$

B.1 Chernoff bounds

Chernoff bounds on the tails of the probability distribution (B.2) are easily obtained from the moment-generating function (B.6). Substitution into (A.5) and (A.7) gives

$$P(U \geq \alpha) \leq \exp[\psi(s)]; \quad s \geq 0 \tag{B.7}$$

and

$$P(U \leq \alpha) \leq \exp[\psi(s)]; \quad s \leq 0 \tag{B.8}$$

where

$$\psi(s) = m(e^s - 1) + s^2\sigma^2/2 - s\alpha \tag{B.9}$$

The optimum value $s = \beta_0$, resulting in the tightest bounds, is obtained by setting the derivative of $\psi(s)$ equal to zero

$$\psi'(\beta_0) = me^{\beta_0} + \sigma^2\beta_0 - \alpha = 0 \tag{B.10}$$

The solution of (B.10) has to be done numerically. A convenient method is the Newton-Raphson algorithm.

The solution β_0 of (B.10) is positive when $\alpha > m$, which correponds to the upper tail bound (B.7), and is negative for $\alpha < m$ corresponding to the lower tail bound (B.8).

The bounds are illustrated in Fig 12 together with the exact probabilites obtained by numerical integration of the saddlepoint integral of Appendix C using the trapezoidal rule. A comparison shows that the ordinary Chernoff bounds are not particulary tight.

For $\sigma = 0$ the bounds (A.10) and (A.11) for a Poisson distribution are obtained and $m = 0$ gives the Chernoff bound for a Gaussian variable

$$P(X \geq \alpha) \leq \exp(-\alpha^2/2\sigma^2) \tag{B.11}$$

B.2 Improved Chernoff bounds

In [9] bounds are presented for the sum of Gaussian noise and arbitrary interference. By a method very similar to the saddlepoint approximation of Appendix C, the following expression is obtained

$$P(U \geq \alpha) \leq \frac{1}{\sqrt{2\pi}|s|\sigma} \exp(s^2\sigma^2/2 - s\alpha)\Psi(s); \quad s > 0 \tag{B.12}$$

where $\Psi(s)$ is the moment-generating function of the stochastic variable representing the interference. Substitution of (B.4) for $\Psi(s)$ in (B.12) gives

$$P(U \geq \alpha) \leq \frac{1}{\sqrt{2\pi}|s|\sigma} \exp[\psi(s)]; \quad s > 0 \tag{B.13}$$

where the function $\psi(s)$ is identical to (B.9)

$$\psi(s) = m(e^s - 1) + s^2\sigma^2/2 - s\alpha \tag{B.14}$$

The bound is optimized by chosing $s = \beta_0$ to minimize $\psi(s)$, i. e. by solving (B.10).

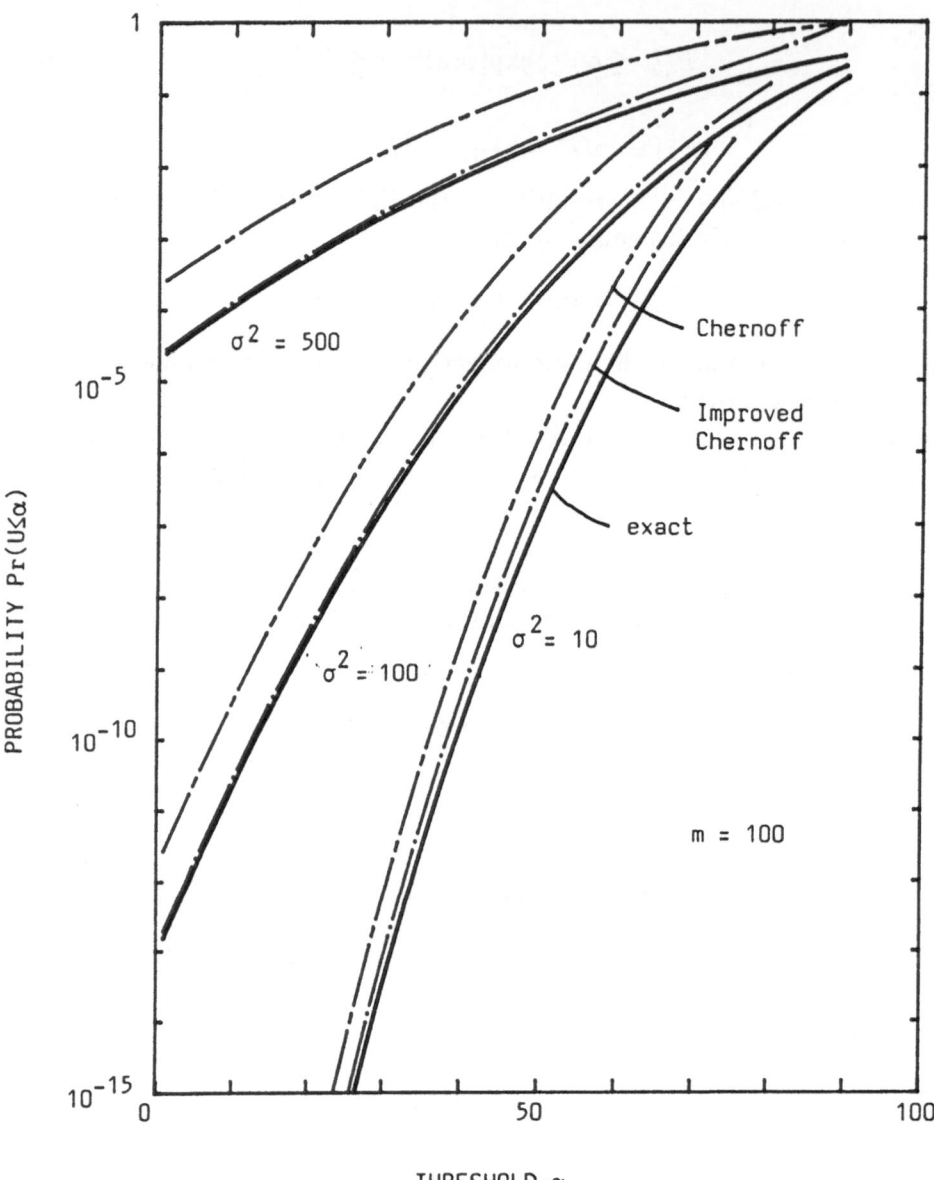

Figure 12:
Cumulative probabilities for the sum of a Poisson variable with
mean $m = 100$ and a Gaussian variable with zero mean and variance σ^2.
Exact value and upper bounds.

a) lower tail.

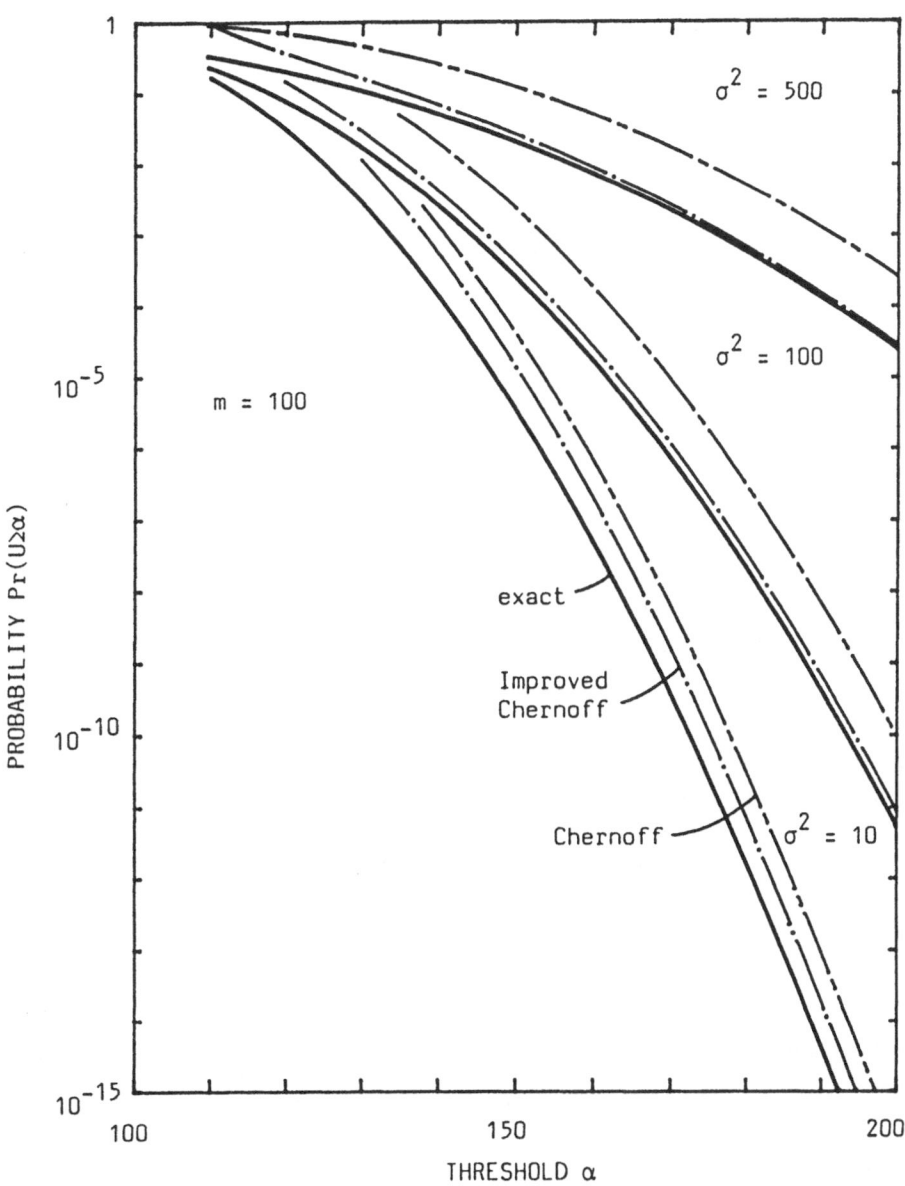

Figure 12:

b) upper tail.

The bound for the lower tail is

$$P(U \le a) \le \frac{1}{\sqrt{2\pi}|\beta_0|\sigma} \exp[\psi(\beta_0)] \tag{B.15}$$

The ordinary Chernoff bounds and the modified bounds (B.13) and (B.15) have the same exponent. The factor $1/\sqrt{2\pi}|\beta_0|\sigma$ is, except for very small values of σ, less than unity which results in an improvement over the ordinary bound. The improved bounds are illustrated in Fig 12.

For $m = 0$ a well-known upper bound for a Gaussian stochastic variable [11] is obtained

$$P(X \ge a) \le \frac{1}{\sqrt{2\pi}(a/\sigma)} \exp(-a^2/2\sigma^2) \tag{B.16}$$

B.3   Saddlepoint approximation

The saddlepoint method described in Appendix C makes exact calculation of the probabilities $P(U \ge a)$ and $P(U \le a)$ possible by numerical quadrature of the integrals (C.5) and (C.11) in Appendix C. Approximate but accurate estimates of the tails of the stochastic variable U can be obtained from the Taylor's series expansion presented in Appendix C.

From (C.8) it follows that the upper tail is approximately equal to

$$P(U \ge a) \approx [2\pi\psi''(\beta_0)]^{-1/2} \exp[\psi(\beta_0)] \tag{B.17}$$

Substitution of $\Psi_u(s)$ from (B.6) into (C.6) gives

$$\psi(s) = m(e^s - 1) + s^2\sigma^2/2 - s\alpha - \ln|s| \tag{B.18}$$

The parameter β_0 is the positive root to the equation

$$\psi'(\beta_0) = me^{\beta_0} + \sigma^2\beta_0 - \alpha - 1/\beta_0 = 0 \tag{B.19}$$

The second derivative of $\psi(\beta)$ is

$$\psi''(\beta_0) = me^{\beta_0} + \sigma^2 + 1/\beta_0^2 \tag{B.20}$$

The lower tail is approximately equal to

$$P(U \le a) \approx [2\pi\psi''(\beta_0)]^{-1/2} \exp[\psi(\beta_0)] \tag{B.21}$$

with β_0 the negative root of (B.19).

The Chernoff bounds and the saddlepoint approximation are similar. The exponents are identical except for the term $\ln|s|$ in (B.18). For $\alpha \gg 1$ the parameter β_0 approaches infinity making the exponents asymptotically the same.

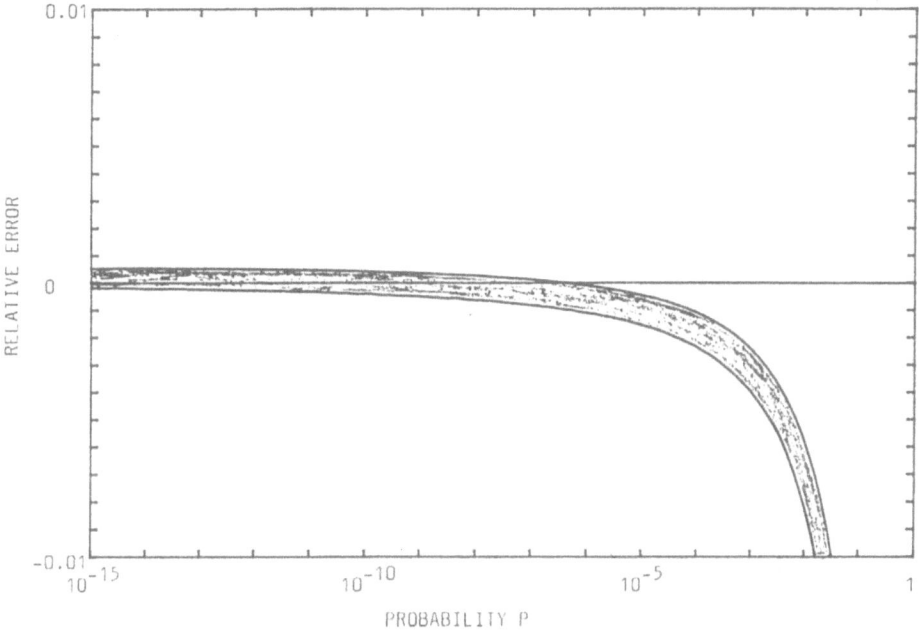

Figure 13:

Accuracy of the saddlepoint approximation of the sum of a Poisson and a Gaussian variable. The relative error $(P - P_{exact})/P_{exact}$ is within the shaded area for $\sigma^2 \geq 10$ for both the upper and the lower tails.

The saddlepoint approximation gives accurate estimates of the tail probabilities. This is illustrated in Fig 13 which shows the relative error for $m = 100$ and σ^2 greater than 10. Both upper and lower tails are included. For probabilities less than 10^{-3} the acurracy is better than 0.4 percent. The saddlepoint approximation is of the same complexity as the improved Chernoff bound but it gives a much more accurate estimate of the true probability.

B.4   Gaussian and Poisson approximations

The compound variable U has mean m and variance $m + \sigma^2$. A widely used approximation is to represent it by a Gaussian distribution with the same mean and variance.

Fig 14 shows the upper and lower tails of of this approximation for $m = 100$ and various values of σ. As expected the approximation becomes more accurate when σ is large compared with m.

An alternative approximation is to represent U by a Poisson distribution with variance (and mean) $m + \sigma^2$. As shown in Fig 14 this approach is best when σ is small. The threshold values in Fig 14 are compensated for the "bias" in the mean introduced by σ^2 for the Poissson case.

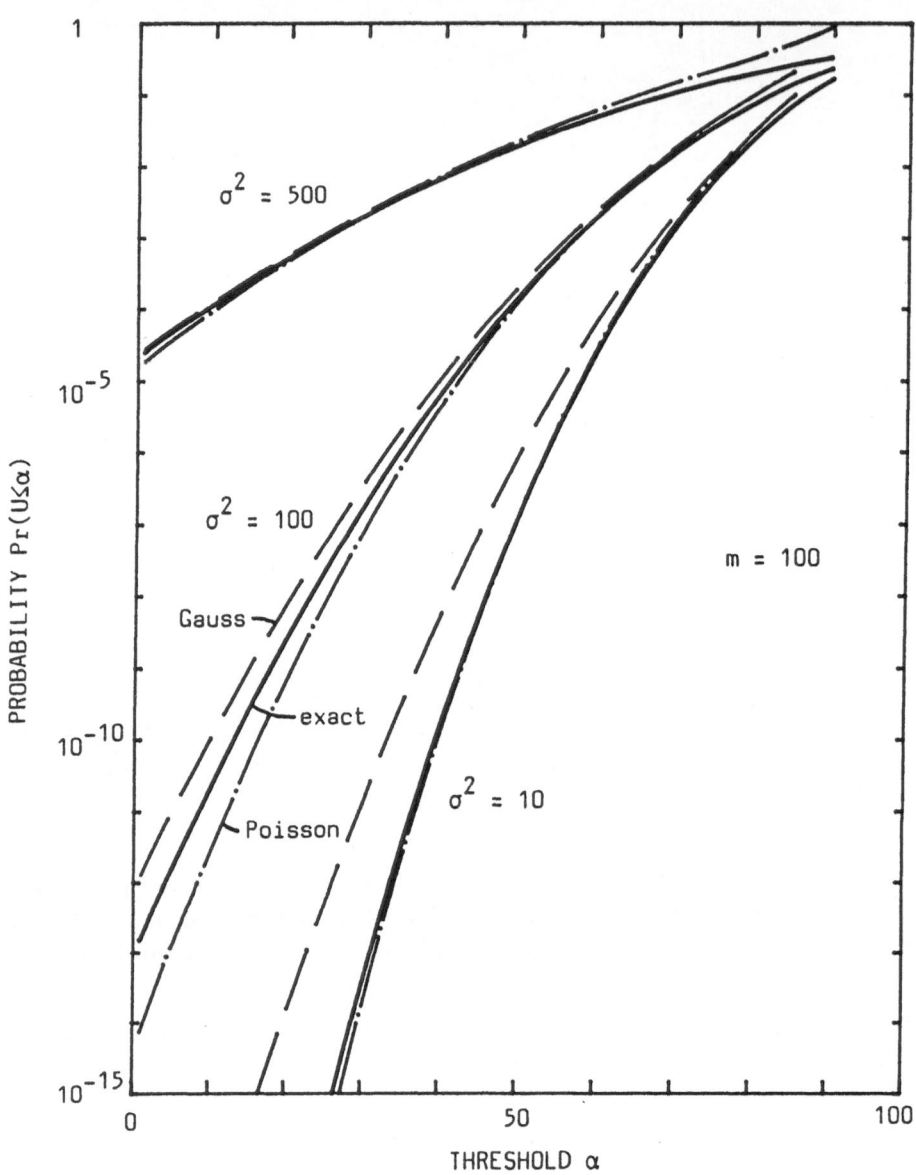

Figure 14:

Gaussian and Poisson approximations.

The upper and the lower tails of the sum of a Poisson and a Gaussian variable compared with Gaussian and Poisson distributions with the same variance.

a) lower tail.

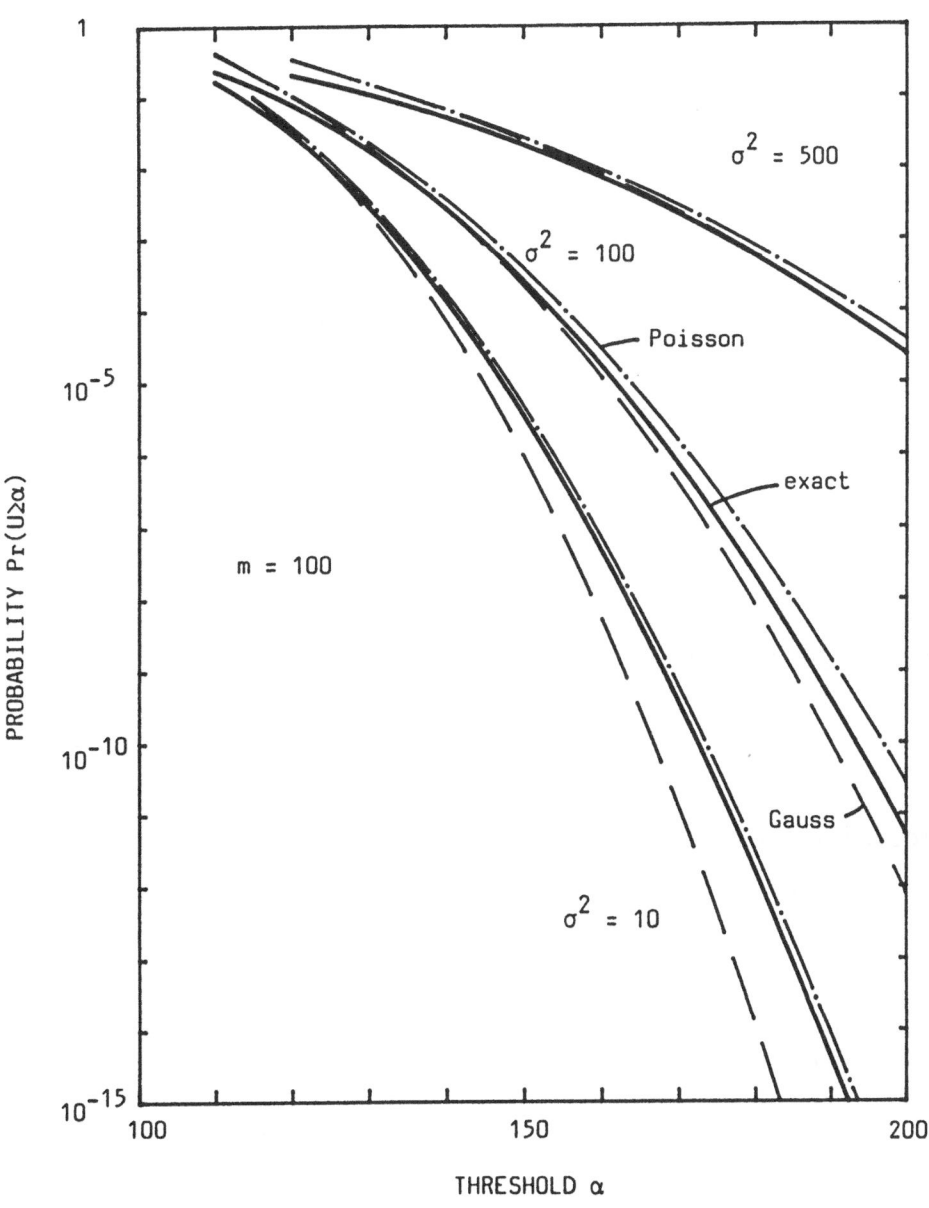

Figure 14:

b) upper tail.

APPENDIX C

SADDLEPOINT APPROXIMATION

C.1   Continous distribution

Following [4] we derive a numerically simple approximation to the cumulative probability distribution of a continous stochastic variable with density function $p(x)$.

Let $q_+(\alpha)$ denote the upper tail

$$q_+(\alpha) = \int_\alpha^\infty p(x)dx \tag{C.1}$$

and

$$q_-(\alpha) = \int_{-\infty}^\alpha p(x)dx \tag{C.2}$$

the lower tail of the probability distribution.

The bilateral Laplace transform of $p(x)$ can be expressed in term of the moment-generating function as

$$\int_{-\infty}^\infty p(x)e^{-sx}dx = \Psi(-s) \tag{C.3}$$

The probability density $p(x)$ is equal to the inverse integral

$$p(x) = \frac{1}{2\pi j} \int_{c-j\infty}^{c+j\infty} \Psi(-s)e^{sx}ds \tag{C.4}$$

where c is in the convergence region of the transform.

Replace p(x) in (C.1) by (C.4) and chose the contour of integration such that $c < 0$ to guarantee convergence of the integral. The result is

$$q_+(\alpha) = \frac{-1}{2\pi j} \int_{c-j\infty}^{c+j\infty} \frac{e^{s\alpha}}{s} \Psi(-s)ds$$

Changing the integration variable from $-s$ to s gives

$$q_+(\alpha) = \frac{1}{2\pi j} \int_{c-j\infty}^{c+j\infty} \frac{e^{-s\alpha}}{s} \Psi(s)ds; \quad c > 0 \tag{C.5}$$

The parameter c is next chosen to be the value of s for which the integrand is minimal. It turns out that this point $s = \beta_0$ corresponds to a saddle point in the complex plane, whence the name of the method.

The integrand is expressed in terms of a "phase" function $\psi(s)$

$$|s|^{-1} \exp(-s\alpha)\Psi(s) = \exp[\psi(s)] \tag{C.6}$$

and $\psi(s)$ is expanded in a Taylor's series around the point $s = \beta_0$

$$\psi(s) = \psi(\beta_0) + \frac{1}{2}\,\psi''(\beta_0)(s - \beta_0)^2 + \cdots \tag{C.7}$$

The first derivative does not appear since $s = \beta_0$ is an extremum of $\psi(s)$.

Substitution of (C.7) into (C.5) neglecting higher-order terms yields the saddlepoint approximation

$$
\begin{aligned}
q_+(\alpha) &\approx \frac{1}{2\pi}\,\exp[\psi(\beta_0)] \int_{-\infty}^{\infty} \exp\left[-\frac{1}{2}\,\psi''(\beta_0)y^2\right] dy \tag{C.8} \\
&= [2\pi\psi''(\beta_0)]^{-1/2}\,\exp[\psi(\beta_0)]
\end{aligned}
$$

The parameter β is the value of s for which $\psi(s)$ has a minimum. It is equal to the positive root of the equation

$$\psi'(\beta_0) = 0 \tag{C.9}$$

For the lower tail, analogously to (C.5),

$$q_-(\alpha) = \frac{-1}{2\pi j} \int_{c-j\infty}^{c+j\infty} \frac{e^{-s\alpha}}{s}\,\Psi(s)ds \tag{C.10}$$

with $c < 0$. Expansion of $\psi(s)$ in a Taylor's series and integration gives

$$q_-(\alpha) \approx [2\pi\psi''(\beta_0)]^{-1/2}\,\exp[\psi(\beta_0)] \tag{C.11}$$

with β_0 equal to the negative root of (C.9).

Only two terms in the Taylor's expansion are used in the derivation above. The approximation can be improved by including higher-order terms [4].

C.2 Discrete distribution

Let P_n; $n = 0, 1, \ldots$ be the probabilities of a discrete stochastic variable. Its generating function is

$$\Phi(z) = \sum_{n=0}^{\infty} z^n P_n \tag{C.12}$$

We define the upper tail of the distribution as

$$Q_+(\alpha) = \sum_{n=\alpha}^{\infty} P_n \tag{C.13}$$

and the lower tail as

$$Q_-(\alpha) = \sum_{n=0}^{\alpha-1} P_n \tag{C.14}$$

The function $\Phi(z)$ can be interpreted as a z-transform with the inverse

$$P_n = \frac{1}{2\pi j} \oint z^{-n-1}\Phi(z)dz \tag{C.15}$$

Summation of (C.15) gives

$$
\begin{aligned}
Q_+(\alpha) &= \frac{1}{2\pi j} \oint \sum_\alpha^\infty z^{-n-1}\Phi(z)dz \\
&= \frac{1}{2\pi j} \oint_{C_1} \frac{z^{-\alpha}}{z-1}\Phi(z)dz \tag{C.16}
\end{aligned}
$$

where C_1, the path of integration, is chosen such that $|z^{-1}| < 1$ to ensure that the sum converges.

The path of integration is then changed to become a vertical line in the complex plane passing through the saddlepoint $z = \beta_0$ with $\beta_0 > 1$.

The parameter of β_0 is determined as the value of z for which the integrand

$$\exp[\phi(z)] = \frac{z^{-\alpha}}{|z-1|}\Phi(z) \tag{C.17}$$

is minimal.

Expanding $\phi(z)$ in a Taylor's series around the point $z = \beta_0$ and integrating yields the approximation

$$Q_+(\alpha) \approx [2\pi\phi''(\beta_0)]^{-1/2} \exp[\phi(\beta_0)] \tag{C.18}$$

with $\phi(z)$ defined by (C.17) and $\beta_0 > 1$ a root of the equation

$$\phi'(\beta_0) = 0 \tag{C.19}$$

In the same way for $Q_-(\alpha)$

$$
\begin{aligned}
Q_-(\alpha) &= \frac{1}{2\pi j} \oint \sum_{n=0}^{\alpha-1} z^{-n-1}\Phi(z)dz \\
&= \frac{1}{2\pi j} \oint_{C_2} \frac{1-z^{-\alpha}}{z-1}\Phi(z)dz \tag{C.20}
\end{aligned}
$$

The integral

$$\oint_{C_2} \frac{1}{z-1}\Phi(z)dz \tag{C.21}$$

is equal to zero if C_2 is chosen such that $|z| < 1$ which gives

$$Q_-(\alpha) = \frac{-1}{2\pi j} \oint_{C_2} \frac{z^{-\alpha}}{z-1}\Phi(z)dz \tag{C.22}$$

The approximation obtained after a Taylor's series expansion and integration is

$$Q_-(\alpha) \approx [2\pi\phi''(\beta_0)]^{-1/2} \exp[\phi(\beta_0)] \tag{C.23}$$

with $\phi(z)$ defined by (C.17) and $0 < \beta_0 < 1$ a root of (C.19).

References

[1] M. Abramowitz and I. A. Stegun
Handbook of Mathematical Functions
National Bureau of Standards,
Applied Mathematics Series, No 55, 1968

[2] D. R. Cox and V. Isham
Point Processes
London: Chapman and Hall, 1980

[3] R. G. Gallager
Information Theory and Reliable Communication
New York: John Wiley and Sons, 1968

[4] C. W. Helstrom
"Approximate evaluation of detection probabilities
in radar and optical communications"
IEEE Trans. Aerospace Electronic Systems
vol. AES-14, pp. 630-640, July 1978

[5] W. M. Hubbard
"The Approximation of a Poisson Distribution
by a Gaussian Distribution"
Proc. IEEE, vol. 58, pp. 1374-1375, Sept. 1970

[6] W. M. Hubbard
"Comparative Performance of Twin-Channnel
and Single-Channel Optical-Frequency Receivers"
IEEE Trans. Commun., vol. COM-20, pp. 1079-1086, Dec. 1972

[7] J. E. Mazo and J. Salz
"On Optical Data Communication via Direct
Detection of Light Pulses
Bell Syst Tech J., vol. 55, pp. 347-369, March 1976

[8] A. Papoulis
Probability, Random Variables, and Stochastic Processes
Second edition
New York: McGraw-Hill, 1984

[9] V. K. Prabhu
"Modified Chernoff Bounds for PAM Systems
with Noise and Interference"
IEEE Trans. Info. Theory, vol. IT-28, pp. 95-100, Jan. 1982

[10] B. Saleh
Photoelectron Statistics
Berlin: Springer-Verlag 1978

[11] J. M. Wozencraft and I. M. Jacobs
Principles of Communication Engineering
New York: John Wiley & Sons, 1965

Bounds on the Size of a Code

Thomas Ericson
Department of Electrical Engineering, Linköping University
S-582 62 Linköping, Sweden

Abstract - Some classical bounds are derived and discussed in a unified manner, along with a few more recent results.

1   Introduction

This paper is tutorial. On the pages to follow we give an overview and a unified treatment of the fundamental existence bounds for block codes. The presentation is restricted to codes over discrete spaces and the characterization is in terms of Hamming metric. Even within these restrictions the treatment is not exhaustive. For instance the linear programming bound and the recent results on algebraic geometry codes are only mentioned, not treated. No preliminary knowledge of coding theory is assumed and all proofs use only elemantary mathematics.

2   The fundamental definitions

Let $\mathcal{A} = \{a_1, a_2, ..., a_q\}$ be an arbitrary finite set. In the present context the set $\mathcal{A}$ will be called an <u>alphabet</u> and the elements a_i of $\mathcal{A}$ will be called <u>letters.</u> Denote by $\mathcal{A}^n$ the set of all n-length sequences $x = x_1 x_2 ... x_n$ of letters from $\mathcal{A}$. We make $\mathcal{A}^n$ into a <u>metric space</u> by defining the distance $d_H(x,y)$ between any two sequences $x = x_1 x_2 ... x_n$ and $y = y_1 y_2 ... y_n$ in $\mathcal{A}^n$ as the number of indices "i" for which $x_i \neq y_i$. The metric d_H is known as the <u>Hamming metric</u>.

Any subset C of $\mathcal{A}^n$ will be referred to as a <u>code</u> and the elements x of C will be called <u>codewords</u>. The parameter n will be called the <u>length</u> of the code and the cardinality $|C|$ will be called the <u>size.</u> In addition we are also interested in the <u>minimum distance</u>,

$$d(C) = \min_{x \neq y} d_H(x,y) ; \qquad x,y \in C .$$

When constructing codes the alphabet $\mathcal{A}$ is often assigned various algebraic properties. In particular, it is quite common to interpret $\mathcal{A}$ as a finite field (which, of course, presupposes that the size of the alphabet is a power of a prime). All such interpretations are made only with the purpose of making various algebraic structures available for the construction. As far as the final characterization of the code is concerned they are of no importance.

Hence, a code C is fully characterized by the quadruple (q,n,M,d), where

q is the size of the alphabet; $q = |\mathcal{A}|$,
n is the length of the code,
M is the size of the code; $M = |C|$,
d is the minimum distance of the code ; $d = d(C)$.

Denote by $C(q,n,M,d)$ the set of all codes C with parameters (q,n,M,d) (over some alphabet $\mathcal{A}$ of size $|\mathcal{A}| = q$). The main problem in coding theory is to determine for which values of the parameters (q,n,M,d) the set $C(q,n,M,d)$ is nonempty. Equivalently the problem is to determine the function

$$A_q(n,d) \triangleq \max\{M: C(q,n,M,d) \neq \emptyset\} .$$

With a few exceptions this function is not known exactly. We proceed to derive some of the known bounds.

3 The sphere packing bound and the Hamming codes

A simple - yet sometimes quite useful - bound on $A_q(n,d)$ is the sphere packing bound. The basic idea is applicable in many different contexts. It simply states that if spheres are packed into a finite space, then the sum of the volumes of the spheres cannot exceed the total volume of the entire space. In order to employ this simple idea in the present context we need the following lemma.

Lemma 1: the codewords x of a code C can be surrounded by non-overlapping t-spheres if and only if $d(C) \geq 2t+1$.

Proof:
For any x the t-sphere $S(x,t)$ surrounding x is defined as

$$S(x,t) \triangleq \{z \in \mathcal{A}^n: d_H(x,z) \leq t\} .$$

Now suppose $d = d(C) \geq 2t + 1$, and let x and y be two distinct codewords in C. Let z be an arbitrary element in the sphere $S(x,t)$. We must show that z is not also in $S(y,t)$. By the triangle inequality we have

$$2t + 1 \leq d_H(x,y) \leq d_H(x,z) + d_H(z,y) .$$

Thus

$$d_H(y,z) = d_H(z,y) \geq 2t + 1 - d_H(x,z) > t ,$$

which shows that z is not an element in $S(y,t)$.

Next suppose $S(x,t) \cap S(y,t) = \emptyset$ for any pair (x,y); $x, y \in C$; $x \neq y$. Fix an arbitrary such pair (x,y) and let $A = \{i_1,i_2,...,i_s\}$ be the set of coordinates "i" for which x and y differ. As $x \notin S(y,t)$ we clearly have $s > t$, and thus we can define $z \in A^n$ according to

$$z_i = \begin{cases} x_i \text{ for } i \in \{i_1,i_2,..., i_t\} \\ y_i \text{ for } i \notin \{i_1,i_2,..., i_t\} \end{cases} .$$

We then have $d_H(y,z) = t$, $d_H(x,z) = s\text{-}t$. As z is in $S(y,t)$ it cannot be in $S(x,t)$ and we conclude $s\text{-}t > t$. Thus $s = d_H(x,y) \geq 2t + 1$.

□

Now combining this lemma with the sphere packing argument gives us the following bound.

Theorem 2: (the sphere packing bound)

Whenever $C(q,n,M,d)$ is nonempty the inequality

$$M \leq M_{SP}(q,n,d) \triangleq \frac{q^n}{\sum\limits_{i=0}^{\left\lfloor \frac{d-1}{2} \right\rfloor} \binom{n}{i}(q-1)^i} ,$$

must be satisfied.

Proof:

We define the volume vol(A) of a set A in $\mathcal{A}^n$ as the number of points in the set: vol(A) $\triangleq$ $|A|$. Then $\text{vol}(\mathcal{A}^n) = q^n$ and $\text{vol}(S(x,t)) = \sum_{i=0}^{t} \binom{n}{i} (q-1)^i$ for each $x \in \mathcal{A}^n$. The theorem follows upon observation of the simple identity

$$\max \{t: 2t + 1 \le d\} = \left\lfloor \frac{d-1}{2} \right\rfloor .$$

□

It follows that $M_{SP}(q,n,d)$ is an upper bound to $A_q(n,d)$. A code C in $\mathcal{A}^n$ with the property that the t-spheres surrounding the codewords x in C fill $\mathcal{A}^n$ precisely for some $t \ge 1$ is said to be perfect. In other words, C is perfect if and only if for some $t \ge 1$ the following two conditions are simultaneously met

i) $x,y \in C, \ x \ne y \Rightarrow S(x,t) \cap S(y,t) = \emptyset$

ii) $\underset{x \in C}{\cup} \ S(x,t) = \mathcal{A}^n .$

Of course this can happen only if d is odd, and then if and only if C fulfils the sphere packing bound with equality, i.e. if and only if $|C| = M_{SP}(q,n,d)$.

Perfect codes do exist. First of all we have the case d = 1, in which the condition $|C| = M_{SP}(q,n,1) = q^n$ is always met. Another example is the case q = 2; n = d odd. In this case we have,

$$\sum_{i=0}^{\left\lfloor \frac{d-1}{2} \right\rfloor} \binom{n}{i}(q-1)^i = \sum_{i=0}^{\frac{n-1}{2}} \binom{n}{i} = \frac{1}{2} \cdot 2^n$$

and consequently $|C| = M_{SP}(2,n,n) = 2$. These two examples are usually referred to as the trivial perfect codes. A class of non-trivial perfect codes is the class of Hamming codes. These can be defined as follows.

Let q be a power of a prime and identify the alphabet $\mathcal{A}$ with the Galois field GF(q). Then $\mathcal{A}^n$ is a linear space with $\mathcal{A}$ = GF(q) as the scalar field. Let $m \ge 1$ be a given integer and let $\{h_i\}_{i=1}^n$ be a set of pairwise linearly independent m-dimensional column vectors. Finally, let n be as large as possible. How large is this? A moment's reflexion reveals that n must be given by

$$n = \frac{q^m - 1}{q - 1}$$

(just notice that $h \in \{h_i\}_{i=1}^{n}$ if and only if $\lambda \cdot h \notin \{h_i\}_{i=1}^{n}$ for $\lambda \neq 1$). Define the m x n matrix H according to

$$H \triangleq [h_1, h_2, ..., h_n]$$

and define the code $C \subseteq GF(q)^n$ as follows:

$$C \triangleq \{x \in GF(q)^n : Hx^T = 0\}.$$

The matrix H has rank m, so C, being the null-space associated with H, has dimension $k = n - m$. It follows that C has size $M = q^{n-m}$. Moreover, if x and y are both in C we have

$$H(x-y)^T = \sum_{i=0}^{n} (x_i - y_i) h_i = 0.$$

But the vectors h_i are pairwise linearly independent. Hence it takes at least three of them to form a linearly dependent set. Thus x and y differ in at least three positions, and we conclude $d(C) \geq 3$. However, the following equality holds

$$\sum_{i=0}^{1} \binom{n}{i} (q-1)^i = 1 + n(q-1) = q^m,$$

which together with the fact $M = q^{n-m}$ shows that $d(C)$ is exactly 3 and the code is perfect.

Thus, the sphere-packing bound is tight - and consequently $A_q(n,d) = M_{SP}(n,d)$ - in the following cases

i) $(q,n,d) = (q,n,1)$; (q,n) arbitrary

ii) $(q,n,d) = (2,n,n)$; n odd

iii) $(q,n,d) = (q, \frac{q^m - 1}{q - 1}, 3)$; q a power of a prime; m = 1,2...

In addition to these three classes of codes there are two isolated cases of perfect codes, namely the Golay codes, for which we have

$$(q,n,d) = (2,23,7)$$
$$(q,n,d) = (3,11,5)$$

respectively. It has been shown that as long as q is a power of a prime there are no other perfect codes (for a full discussion we refer to MacWilliams - Sloane [1], Ch. 6:8). Hence, for quite many values of the parameters (q,n,d) the sphere packing bound is known not to be tight. This calls for further bounds.

4   The Singleton bound and the Reed-Solomon codes

Theorem 3: (the Singleton bound)

Whenever C(q,n,M,d) is nonempty, the inequality

$$M \leq M_S(q,n,d) \triangleq q^{n-d+1}$$

must be satisfied.

Proof:
Let $C \in C(q,n,M,d)$ and choose k such that $q^{k-1} < M \leq q^k$. Consider any k-1 coordinates of the codewords, say i=1,2,..., k-1. As $M > q^{k-1}$ there must be two codewords, say x,y $\in$ C, such that $x_i = y_i$; i=1,2,... k-1. It follows that $d \leq d_H(x,y) \leq n-k+1$, and we conclude

$$M \leq q^k \leq q^{n-d+1} \quad .$$

Codes satisfying the Singleton bound with equality are said to be maximum distance separable, or MDS for short. Once again we have an important class of codes meeting this criterion, namely the so called Reed-Solomon codes (RS codes for short). These can be defined in many different ways. We prefer the following definition.

Again let $\mathcal{A}$ be identified with the Galois field GF(q) and let $a_0, a_1,...$, a_{n-1} be different elements i GF(q) (this implies $n \leq q$), and define the matrix H as follows

$$H = \begin{bmatrix} 1 & 1 & & 1 \\ a_0 & a_1 & & a_{n-1} \\ & & & \\ & & & \\ a_0^{r-1} & a_1^{r-1} & & a_{n-1}^{r-1} \end{bmatrix} .$$

The Reed-Solomon code C is defined as

$$C \triangleq \{x \in GF(q)^n : Hx^T = 0\} \ .$$

It is clear that the rank of H is at most r, so the dimension k of the code C satisfies $k \geq$ n-r. Moreover, any r columns from H form a Vandermonde matrix, which is nonsingular because of the assumption that the a_i's are different. It follows that it takes at least r+1 columns from H to form a linearly dependent set, and by the same argument as for the Hamming codes we conclude that the minimum distance d = d(C) satisfies $d \geq$ r+1. Combining this estimate with the above estimate for the dimension k we get

$$k \geq n\text{-}d\text{+}1$$

or, what is the same,

$$M = q^k \geq q^{n\text{-}d\text{+}1} \ .$$

By the Singleton bound equality actually prevails and the code is MDS.

By this construction we have MDS codes for all parameters (q,n,d) as long as q is a power of a prime and n satisfies $n \leq q$. Hence the Singleton bound is tight in all those cases. In fact, the condition $n \leq q$ can be slightly relaxed, but precise conditions for tightness of the bound are not known. There are, however, many cases where it is known not to be tight. For instance, the Hamming codes (which always meet the sphere-packing bound) are MDS only for m = 2.

5   Constant weight codes and the Johnson bounds

Neither the sphere packing bound nor the Singleton bound are tight except in a few specific cases. In particular they are often quite weak for small values of q. For that reason we now develop some additional bounds for binary codes (q = 2). We consider first the special case of constant weight codes. These codes are of interest in their own right, but the bounds we get can also be used to generate bounds for codes in general.

Without loss of generality, let the alphabet be $A = \{0,1\}$. By the weight $w_H(x)$ (Hamming weight) of a binary sequence $x = x_1 x_2 \ldots x_n$ we understand the number of occurrences of the letter "1" in the sequence. A constant weight code C is a subset of $\{0,1\}^n$ with the property that all codewords x in C have the same weight w:

$$w_H(x) = w ; \ x \in C .$$

We define the <u>correlation</u> $c(x,y)$ between two sequences $x,y \in \{0,1\}^n$ as the number of coordinates "i" for which $x_i=y_i=1$,

$$c(x,y) \triangleq \sum_{i=1}^{n} x_i y_i .$$

We notice the identity

$$d_H(x,y) = w_H(x) + w_H(y) - 2c(x,y) ,$$

which in the special case $w_H(x) = w_H(y) = w$ takes the form

$$d_H(x,y) = 2w - 2c(x,y) .$$

It follows that all distances in a constant weight code are <u>even.</u> The maximum correlation $c(C)$ of a constant weight code C is defined by

$$c(C) \triangleq \max_{x \neq y} c(x,y) ; \ x,y \in C .$$

For constant weight codes we find this quantity more convenient to work with than the minimum distance.

Denote by $CW(n,w,c,T)$ the set of all constant weight codes C of length n, weight w, maximum correlation c, and size T (a constant weight code is always binary, so q need not be specified). We are interested in determining for what values of the parameters (n,w,c,T) the set $CW(n,w,c,T)$ is nonempty. Equivalently we want to determine the function

$$T(n,w,c) \triangleq \max\{T: CW(n,w,c,T) \neq \emptyset \} .$$

The following bounds apply.

Theorem 4: (the Johnson bounds)

i) $T(n,w,c) \leq \frac{n}{w} T(n-1,w-1,c-1)$ (always)

ii) $T(n,w,c) \leq n \cdot \frac{w-c}{w^2-nc}$ if $w^2 > nc$.

Proof:

Let $T = T(n,w,c)$ and suppose $C \in CW(n,w,c,T)$. Let $X = \{x_{ij}\}$ be the codeword matrix (i.e. x_{ij} is the j-th component of the i-th codeword; $i=1,2...$, T; $j=1,2,...,n$). Define $k_v \triangleq \sum_{i=1}^{T} x_{iv}$, i.e. let k_v denote the number of codewords x in C which have a "1" in the v-th coordinate. We notice the identity

$$\sum_{v=1}^{n} k_v = wT$$

and the inequality

$$k_v \leq T(n-1,w-1,c-1)$$

(the inequality follows by observing that the codewords x in C with a "1" in the v-th position generate a code $C' \in CW(n-1,w-1,c-1,k_v)$ upon deletion of the v-th coordinate). The first bound now follows right away, while the second bound follows if we combine the simple inequalities

$$\sum_{i \neq j} \sum_{v=1}^{n} x_{iv} x_{jv} = \sum_{i \neq j} c(x_i, x_j) \leq T(T-1) c$$

and

$$\sum_{v=1}^{n} \sum_{i \neq j} x_{iv} x_{jv} = \sum_{v=1}^{n} k_v(k_v-1) \geq \frac{1}{n} w^2 T^2 - wT .$$

$\square$

The following corollary is immediate

Corollary 5:

$$T(n,w,c) \leq \left\lfloor \frac{n}{w} \left\lfloor \frac{n-1}{w-1} \cdots \left\lfloor \frac{n-c}{w-c} \right\rfloor \cdots \right\rfloor \right\rfloor$$

$$\leq J(n,w,c) \triangleq \prod_{i=0}^{c} \frac{n-i}{w-i} = \frac{\binom{n}{c+1}}{\binom{w}{c+1}} .$$

There exist constant weight codes C meeting the bound $J(n,w,c)$. As an example, let C be a binary Hamming code with parameters $n = 2^m-1$, $M = 2^{n-m}$, $d = 3$. Define

$$\mathcal{B} \triangleq \{x \in C : w_H(x) = 3 \}.$$

By construction $\mathcal{B}$ is a constant weight code with weight $w = 3$. The minimum distance is 3 in C ; it certainly cannot be smaller in $\mathcal{B}$. Moreover, $\mathcal{B}$ is a constant weight code, so all distances in $\mathcal{B}$ are even. We conclude $d(\mathcal{B}) \geq 4$, or

$$c = c(\mathcal{B}) = w - \frac{1}{2} d(\mathcal{B}) \leq 3 - \frac{1}{2} \cdot 4 = 1 \ .$$

However, $c(\mathcal{B})$ is actually precisely 1, because $|\mathcal{B}| = J(n,3,1)$. To see this, notice that for each $x \in \mathcal{B}$ the sphere $S(x,1)$ contains precisely 3 vectors z of weight $w_H(z) = 2$. Moreover, each vector z of weight $w_H(z) = 2$ is contained in <u>some</u> sphere $S(x,1)$; $x \in \mathcal{B}$. Thus

$$3 \cdot |\mathcal{B}| = \binom{n}{2}$$

or

$$|\mathcal{B}| = \frac{1}{3}\binom{n}{2} = J(n,3,1) \ .$$

Constant weight codes satisfying the bound $T \leq J(n,w,c)$ with equality are closely related to so called <u>Steiner systems</u>. Let $S = \{s_1,s_2,..., s_n\}$ be an arbitrary finite set. A Steiner system is a collection $\{A_1,A_2,... , A_b\}$ of subsets (usually called <u>blocks</u>), each of size $|A_i| = w$, with the property that any t-subset (i.e. $B \subseteq S$ with $|B| = t$) is contained in precisely <u>one</u> of the sets A_i. As each A_i contains $\binom{w}{t}$ t-subsets and as there are $\binom{n}{t}$ such subsets altogether in S we clearly have

$$b = \frac{\binom{n}{t}}{\binom{w}{t}} = J(n,w,t-1) \ .$$

The collection $\{A_1,A_2... , A_b\}$ is referred to as an $S(n,w,t)$ - system. The connection with constant weight codes is obtained by introducing the <u>incidence matrix</u> $X = (x_{ij})$, where

$$x_{ij} = \begin{cases} 1 & s_j \in A_i \\ 0 & s_j \notin A_i \end{cases} .$$

It is clear that X can also be interpreted as the <u>code matrix</u> of a binary code $\mathcal{B}$ with length n, constant weight w, and maximal correlation c = t-1. The size of the code is

$$|\mathcal{B}| = b = J(n,w,c) .$$

Conversely it is clear that any code $\mathcal{B} \in CW(n,w,c,T)$ with $T = J(n,w,c)$ induces a Steiner system $S(n,w,c+1)$.

Steiner systems were studied long before coding theory was in existence, and quite a number of such systems are known. For an extensive discussion we refer to [3].

It follows that the bound $J(n,w,c)$ is tight in several cases. On the other hand, it is also known <u>not</u> to be tight in many other cases. For instance, the second Johnson bound

$$T(n,w,c) \leq n \cdot \frac{w-c}{w^2-nc}; \quad w^2 > nc$$

is usually tighter when it is valid. Also this bound is sometimes met with equality. The following construction gives an example.

Let H be an m x m <u>Hadamard</u> matrix. This means that H has the form $H = (h_{ij})$, where h_{ij} equals +1 or -1, and

$$H \cdot H^T = mI_m$$

(I_m is the m x m unity matrix). Without loss of generality we may assume that H is <u>normalized</u>, which means that the first row and the first column in H consist of +1's only: $h_{1i} = h_{i1} = 1$; i = 1,2,..., m.

Notice that any pair of rows in H differ in exactly m/2 positions (and consequently also agree in exactly m/2 positions; this implies that m is even; as a matter of fact m is always a multiple of 4 whenever it is larger than 2). Let X be the matrix obtained from H by first deleting the first row and first column (those consisting of +1's only) and then replacing +1 with 0 and -1 with 1. Let C be the binary code having the rows in X as its codewords. It is easy to see that $C \in CW(n,w,c,T)$, with n = T = m-1; w = m/2 ; c = m/4. A simple calculation reveals that

$$n \cdot \frac{w-c}{w^2-nc} = m-1$$

which shows that the second Johnson bound is satisfied with equality for this code.

Several constructions of Hadamard matrices are known. The simplest is the following. Let H_0 be the trivial 1×1 - matrix $H_0 = [1]$. Define $\{H_i\}_{i=1}^{\infty}$ recursively by the equation

$$H_{i+1} = \begin{bmatrix} H_i & H_i \\ H_i & -H_i \end{bmatrix}$$

$$i = 0,1,2,...$$

It is easy to see that H_i is an $2^i \times 2^i$ Hadamard matrix.

6 Two useful inequalities

We now provide two results relating the existence of constant weight codes to the existence of general codes. Of course we have the trivial inclusion $CW(n,w,c,T) \subseteq C(2,n,T,2(w-c))$. The following two results are more interesting as they go in the opposite direction.

Theorem 6: For any w; $0 \le w \le n$:

$$A_2(n,d) \le \frac{2^n}{\binom{n}{w}} T\left(n,w,\left[w - \frac{d}{2}\right]\right) .$$

Proof:
Let $M = A_2(n,d)$ and suppose $C \in C(2,n,M,d)$. Denote $F = \{0,1\}$, and for any $x,y \in F^n$ define

$$G(x,y) = \begin{cases} 1 & d_H(x,y) = w \\ 0 & d_H(x,y) \ne w \end{cases} .$$

For any $x \in C$ we have

$$\sum_{y \in F^n} G(x,y) = \binom{n}{w} ,$$

and so

$$\sum_{x \in C} \sum_{y \in F^n} G(x,y) = |C| \binom{n}{w} \ .$$

Now for any $y \in F^n$ define

$$C_y \triangleq \{x \oplus y: \ x \in C \ ; \ d_H(x,y) = w\}\cdot$$

where "$\oplus$" denotes addition modulo 2.

It is clear that C_y is a constant weight code with weight w and maximum correlation $c \le w - d/2$.

Thus

$$|C_y| = \sum_{x \in C} G(x,y) \le T(n,w, \lfloor w\text{-}d/2 \rfloor)$$

and so

$$\sum_{y \in F^n} \sum_{x \in C} G(x,y) \le 2^n \cdot T(n,w,\lfloor w\text{-}d/2 \rfloor) \ .$$

$\square$

This result will be used in section 8 when we derive the Bassalygo-Elias bound.

Theorem 7: $A_q(n,d) \le T(qn,n,n\text{-}d)$.

Proof:
Let $C \in C(q,n,M,d)$, where $M = A_q(n,d)$, and represent the alphabet $\mathcal{A} = \{a_1, a_2,..., a_q\}$ by binary sequences of length q and weight 1. The resulting binary code $\tilde{C}$ generated in this way by C has length qn, weight n and correlation $c = n\text{-}d$. The size is of course $|\tilde{C}| = |C| = M$.

$\square$

Combining this result with the second Johnson bound gives us the following bound.

Theorem 8: (the Plotkin bound)

Whenever $d > \frac{q-1}{q} \cdot n$ the following bound applies:

$$A_q(n,d) \leq M_P(q,n,d) \triangleq \frac{qd}{qd-(q-1)n} \cdot$$

Like all our previous bounds the Plotkin bound is also sometimes satisfied with equality. In the binary case (q=2) an example is obtained by a minor modification of our example at the end of section 5. If we let X be the codeword matrix obtained by deleting only the first <u>column</u> in H (and not the first <u>row</u>) we get a binary code with parameters n=m-1, M=m, d=m/2. By a simple calculation we see that the Plotkin bound is satisfied with equality. Taking from this code only those codewords with a "0" in the first position, and then deleting this first zero, produces another code with parameters n=m-2; M = m/2; d=m/2. Again the Plotkin bound is met with equality.

7   The Gilbert-Varshamov bound

All the bounds we have considered so far are upper bounds to the size of a code. Although in each case we gave explicit examples of codes satisfying the bounds with equality, all those constructions exist only for certain specific values of the code parameters, while the <u>bounds</u> of course are valid for <u>all</u> values of the parameters. The fundamental result as far as <u>lower bounds</u> are concerned is the Gilbert-Varshamov bound, discovered independently by Gilbert [4] and Varshamov [5], who gave slightly different versions of the same basic idea. Of course, each specific code provides us with a lower bound to $A_q(m,d)$ (or T(n,w,c) as the case may be) for <u>some</u> values of the parameters. The strength of the Gilbert-Varshamov bound is that it provides us with a bound for <u>any</u> choice of the parameters.   Unfortunately the argument is non-constructive in the sense that the number of steps needed in the construction of explicit codes using the Gilbert-Varshamov idea increases exponentially with the codelength.

The arguments given by Gilbert and Varshamov are somewhat different, and even the exact form of their results differ. Gilbert's argument is purely combinatorial and more general, although somewhat weaker. Varshamov's argument applies to <u>linear</u> codes. This is stronger, as it further limits the class of codes for which the result holds, and also because linear codes are of particular interest in applications. In the present discussion both versions will be presented, along with a modification of Varshamov's result obtained by Blokh-Zyablov [6].

In any metric space S with distance function d denote by $V(r)$ the volume of a sphere with radius r:

$$V(r) \triangleq \text{vol}(S(x,r)) ; \quad x \in S ;$$

(we assume that $\text{vol}(S(x,r))$ is indepedet of x). Although the argument could easily be generalized our interest is limited to discrete spaces. We therefore restrict $d(x,y)$ to take integer values only, and as above we define the volume of a set A as the number of points in the set:

$$\text{vol}(A) = |A| .$$

Let $A(d)$ be the maximal number of points which can be selected from S in such a way that all points are separated by distance at least d. Gilbert's bound can be stated as follows.

Theorem 9: (Gilbert's bound)

$$A(d) \geq \frac{\text{vol}(S)}{V(d-1)} .$$

Proof:
Define for each pair (x,y) of points in S the function $F(x,y)$ as follows:

$$F(x,y) = \begin{cases} 1 & d(x,y) < d \\ 0 & d(x,y) \geq d \end{cases} .$$

Let $C \subseteq S$ satisfy $|C| = A(d)$; $d(C) \geq d$. For any $x \in C$ we have

$$\sum_{y \in S} F(x,y) = \text{vol}(S(x,d-1)) = V(d-1) ,$$

and so

$$\sum_{x \in C} \sum_{y \in S} F(x,y) = |C| \cdot V(d-1) .$$

On the other hand, for each $y \in S$ there must be some $x \in C$ such that $d(x,y) \leq d-1$, because in the opposite case, i.e. if we could find some y such that $d(x,y) \geq d$ for all $x \in C$, it is clear that C could be enlarged by y without violating the condition $d(C) \geq d$. This, however, contradicts the assumption $|C| = A(d)$. Consequently

$$\sum_{x \in C} F(x,y) \geq 1$$

for all $y \in S$, and so

$$\sum_{y \in S} \sum_{x \in C} F(x,y) \geq \text{vol}(S) = |S| .$$

□

Taking S as the space $\mathcal{A}^n$ for some alphabet $\mathcal{A}$ of size $|\mathcal{A}| = q$ and choosing $d = d_H$ we get (cf Theorem 2)

$$V(d-1) = \sum_{i=0}^{d-1} \binom{n}{i}(q-1)^i ;$$

which gives the following corollary.

Corally 10 (Gilbert, 1952): For any $n \geq d \geq 1$

$$A_q(n,d) \geq A_G(n,d) \triangleq \frac{q^n}{\sum_{i=0}^{d-1} \binom{n}{i}(q-1)^i} .$$

As another special case, let S be the set of all binary sequences of weight w . Then $\text{vol}(S) = \binom{n}{w}$, and the distance (we still assume Hamming distance d_H) is always an even number. Let the minimum distance be $d=2l$. A simple calculation reveals

$$V(d-1) = \sum_{i=0}^{l-1} \binom{w}{i}\binom{n-w}{i} ,$$

and by using the relation $l = w-c$ we get

Corollary 11: For any $n \geq w > c \geq 0$

$$T(n,w,c) \geq \frac{\binom{n}{w}}{\sum_{i=0}^{w-c-1} \binom{w}{i}\binom{n-w}{i}} .$$

When deriving upper bounds we make as few assumptions as possible in order to get the strongest possible results. When deriving lower bounds the situation is opposite: various restrictions only improve the result; at best a single code is specified. Thus it is an interesting result that essentially the same bound can be obtained even if we restrict ourselves to so called linear codes. A definition follows.

We have already seen that when q is a power of a prime it is often convenient to associate the alphabet $\mathcal{A}$ with the finite field GF(q). A code C is then a subset of the linear space of all n-length sequences of symbols from $\mathcal{A}$ = GF(q). A linear code is nothing else than a linear subspace in $GF(q)^n$. The size of a linear code is of course always of the form $M = q^k$, where k is the dimension of the code.

A convenient way of specifying a linear code is to define it as the null-space of a certain matrix H called the parity check matrix. (This idea has already been used above when defining the Hamming codes and the Reed-Solomon codes). The following lemma provides the crucial characterization.

Lemma 12: let the code C be defined according to

$$C \triangleq \{x \in GF(q)^n : Hx^T = 0\} ,$$

where H is an m x n matrix over GF(q). Then

dim C = n - rank H
$d(C)$ = smallest number of linearly dependent columns in H.

The rank r of H is often called the redundancy of the code C. This refers to the fact that r is the number of extra symbols added in the representation of each message in order to achieve the desired minimum distance. Notice that while d = $d(C)$ is the smallest number of linearly dependent columns the redundancy r = rank H is the largest number of linearly independent columns in H. While it is desirable to have d large we generally desire r to be as small as possible. The linear coding problem can be viewed as finding matrices H with a favourable trade-off between these two features.

Theorem 13: (Varshamov 1957)

For any $n \geq d \geq 2$ there exists a linear code $C \subseteq GF(q)^n$ such that

$$|C| \geq A_V(n,d) \triangleq \frac{q^{n-1}}{\sum\limits_{i=0}^{d-2} \binom{n-1}{i}(q-1)^i} .$$

Proof:

Let us denote

$$H = [h_1, h_2, ..., h_n] \; ; \quad h_i \in GF(q)^r .$$

The problem is to specify the columns h_i such that any d-1 of them are linearly independent.

Clearly all columns are non-zero. Let us choose them successively. The first column h_1 is arbitrary. Let us consider the situation when s-1 columns have been successfully chosen. The problem is to choose h_s such that h_s is different from all non-zero linear combinations that can be formed from d-2 or fewer of the previously selected columns $h_1, h_2, ..., h_{s-1}$. There are at most

$$\sum_{i=1}^{d-2} \binom{s-1}{i}(q-1)^i$$

of those, and clearly h_s can be chosen different from all of them as long as this number is strictly less than q^r-1 (the total number of different non-zero vectors in $GF(q)^r$). Obviously this criterion becomes increasingly severe as s increases, and the crucial step is of course the last one, s = n. Thus we see that H can be constructed so as to specify a linear code with minimum distance d and redundancy rank H $\leq r$ as long as the following condition is satisfied

$$\sum_{i=0}^{d-2} \binom{n-1}{i}(q-i)^i < q^r .$$

In order to obtain as good a code as possible we choose for given n,d, the parameter r as small as possible, whereby

$$q^{r-1} \leq \sum_{i=0}^{d-2} \binom{n-1}{i}(q-1)^i .$$

Combining this last inequality with lemma 12 gives us the desired result.

□

A simple calculation reveals that Varshamov's bound is always larger than Gilbert's bound. This is remarkable, as Varshamov's result applies to a restricted class of codes - linear codes - while no such restriction applies to Gilbert's result. Both bounds are, however, quite weak for small values of r; most explicit constructions give better results in those cases. The importance of the bounds by Gilbert and Varshamov is in the asymptotic case (to be treated in section 8), where they coincide.

Still another, slightly different result has been derived by Blokh-Zyablov. They showed that it is even possible to have a sequence of nested linear codes, all satisfying a bound of the Gilbert-Varshamov type. This fact is of interest in the study of concatenated codes, [6]. The result is as follows.

Theorem 14: (Blokh-Zyablov, 1982)

There exists a sequence $\{C_k\}_{k=1}^n$ of nested linear codes:

$$C_1 \subseteq C_2 \subseteq \ldots \subseteq C_n$$

such that

$$|C_k| = q^k \geq A_{BZ}(n,d_k) \triangleq \frac{q^{n+1}}{\sum\limits_{i=0}^{d_k} \binom{n}{i}(q-1)^i} .$$

where $d_k = d(C_k)$; $k = 1,2,\ldots, n$.

Proof:
The proof goes by induction. Let c_1 be an arbitrary vector in $GF(q)^n$ such that all n components in c_1 are non-zero. Let $C_1 = \{\lambda c_1 : \lambda \in GF(q)\}$. Clearly $|C_1| = q$; $d_1 = d(C_1) = n$, so C_1 satisfies the bound with equality. Now suppose $C_1, C_2\ldots , C_k$ have already been chosen in accordance with the theorem. Denote

$$V(d) \triangleq \sum\limits_{i=0}^{d} \binom{n}{i}(q-1)^i$$

and notice that the theorem assures the following relations:

$$|C_k| = q^k$$

$$V(d_k) \geq q^{n-k+1}$$

$$k = 1,2,..., n \; .$$

Our induction hypothesis is that these relations are fulfilled for all the codes $C_1, C_2,.... ,$ C_k. We now make use of a familiar concept for linear codes known as the <u>standard array</u>. Regard C_k as a subgroup in the linear group $GF(q)^n$ and let $\mathcal{B}_i$; $i = 1,2,... , q^{n-k}$, be the corresponding cosets. From each coset $\mathcal{B}_i$ choose an element e_i of minimum weight and consider the set $\mathcal{E} = \{e_i\}_{i=1}^{q^{n-k}}$ (the elements e_i are usually called "coset leaders"). The <u>standard array</u> is simply an arrangement of the elements in $GF(q)^n$ in an $q^k \times q^{n-k}$ matrix, with the rows labeled by elements e from $\mathcal{E}$ and the columns labeled by elements x from C_k.

Now from the set $\mathcal{E}$ choose an element c_k of maximum weight, let $w_k \triangleq w_H (c_k)$ and let C_{k+1} be the linear space spanned by C_k and c_k. It follows that $|C_{k+1}| = q^{k+1}$, and that $d_{k+1} = \min(d_k, w_k)$. We notice that $V(w_k) \geq q^{n-k}$, with equality if and only if C_k is perfect. We already know, by the induction hypothesis, that $V(d_k) \geq q^{n-k+1} > q^{n-k}$. Thus

$$q^{n-k} \leq \min(V(d_k), V(w_k))$$

$$= V(\min(d_k, w_k)) = V(d_{k+1}) \; .$$

$\square$

8   Asymptotic bounds

We conclude by a short discussion of the asymptotic properties of $A_q(n,d)$ and $T(n,w,c)$ as the parameters, (n,d) and (n,w,c) respectively, tend to infinity. This question is interesting in itself from a purely mathematical point of view. However, our interest in this problem is mainly motivated by its importance for the application of coding theory in telecommunication engineering.

Suppose q-ary symbols are to be transmitted over an unreliable channel such that each transmission is subject to an error with probability p. Then the expected number of errors in transmission of n symbols is np. If n is large we may, under quite general assumptions, rely on the law of large numbers and be sure that the <u>actual</u> number of errors (which of course is a stochastic variable) has a very small probability of exceeding that number by more than a small fraction.

Thus if we transmit a codeword x from a code $C \in C(q,n,M,d)$ with d slightly larger than twice the number np, then clearly with very high probability the received sequence will be closer to the transmitted codeword than to any other codeword in C, and as long as we restrict ourselves to transmitting only codewords from C reliable communication can be maintained over the unreliable channel. Clearly, the restriction we impose on ourselves only to transmit codewords from C limits the number of messages we can transmit; so in order for the strategy to be efficient it is essential that M is not too small.

This may be motivation enough for an investigation of the asymptotic behaviour of $A_q(n,d)$ as n and d tend to infinity in such a way that d/n tend to some fixed number δ (clearly δ should satisfy $\delta > 2p$). It should also be clear that we could expect an exponential increase in $A_q(n,d)$ under these conditions. Hence we will be interested in the quantity

$$R(q,\delta) \triangleq \sup \overline{\lim_{n \to \infty}} \; \frac{1}{n} \log A_q(n,d)$$

where the sup is taken over all sequences $\{d_n\}$ such that $\frac{d_n}{n} \to \delta$. Following an established convention we will usually choose the base 2 for the logarithm, although occasionally it might be more convenient to use the base q.

For constant weight codes we will be interested in the quantity

$$E(v,\kappa) \triangleq \sup \overline{\lim_{n \to \infty}} \; \frac{1}{n} \log T(n,w_n,c_n)$$

where the sup is taken over all sequences $\{w_n\}_{n=1}^{\infty}$, $\{c_n\}_{n=1}^{\infty}$, such that

$$\frac{w_n}{n} \to v; \; \frac{c_n}{w_n} \to \kappa \; .$$

Each one of the bounds we have derived in the previous sections gives rise to a bound on the exponent $R(q,\delta)$ or $E(v,\kappa)$ respectively. The key result needed to obtain these bounds is given in the following lemma.

Lemma 15: For any sequence $\{x_n\}_{n=1}^{\infty}$ such that $\frac{x_n}{n} \to x$ we have

$$\lim_{n \to \infty} \frac{1}{n} \log \binom{n}{x_n} = h(x)$$

where h(x) is the following function:

$$h(x) \triangleq -x \log x - (1-x)\log(1-x)$$

(h(x) is known as the <u>binary entropy function</u>).

Proof:

The proof follows by a straightforward application of Stirlings approximation formula for factorials.

□

Combining this lemma with Theorem 2 (the sphere packing bound) we readily get the following bound on $R(q,\delta)$, namely

$$R(q,\delta) \leq R_{SP}(q,\delta) \triangleq \log q - h(\delta/2) - \delta/2 \log(q-1) ;$$

$$0 \leq \delta < 1 .$$

From the Plotkin bound (Theorem 8) we immediately get

$$R(q,\delta) = 0 \; ; \; \frac{q-1}{q} \leq \delta \leq 1 ,$$

so actually the sphere packing bound $R_{SP}(q,\delta)$ is really of interest only in the interval $0 \leq \delta < \frac{q-1}{q}$. The asymptotic form of the Singleton bound (Theorem 3) is $R(q,\delta) \leq (1-\delta) \log q$, but this bound is everywhere weaker than the two previous bounds, and is therefore of very little interest.

To get the asymptotic form of the Gilbert-Varshamov bound we start by observing that the sum

$$V(r) = \sum_{i=0}^{r} \binom{n}{i}(q-1)^i$$

is dominated by its last term as long as

$$q-1 \geq \frac{r}{n-r} \ .$$

Hence we have the bound

$$V(r) \leq \begin{cases} (r+1)\binom{n}{r}(q-1)^r & 0 \leq r \leq n \cdot \frac{q-1}{q} \\ \\ q^n & n \cdot \frac{q-1}{q} \leq r \leq n \ . \end{cases}$$

Combining this bound with Lemma 15 and any one of the bounds in Corollary 10, Theorem 13, or Theorem 14 gives us the bound

$$R(q,\delta) \geq R_{GV}(q,\delta) \triangleq \begin{cases} \log q - h(\delta) - \delta \log(q-1) & 0 \leq \delta \leq \frac{q-1}{q} \\ \\ 0 & \frac{q-1}{q} < \delta \leq 1 \ . \end{cases}$$

It is of interest to note that this bound is positive everywhere except in the range $\frac{q-1}{q} \leq \delta \leq$ 1, where we know from the Plotkin bound that $R(q,\delta)$ is actually zero.

Also notice the close relationship between the sphere packing bound and the Gilbert-Varshamov bound:

$$R_{SP}(q,\delta) = R_{GV}(q,\delta/2) \ ; \quad 0 \leq \delta < \frac{q-1}{q} \ .$$

These bounds are illustrated in fig. 1. If instead of $R(q,\delta)$ we consider the inverse function $\delta(R)$ our bounds take the following nice form:

$$\delta_{VG}(R) \leq \delta(R) \leq \delta_{SP}(R) = 2\delta_{VG}(R)$$

where, of course, δ_{VG} and δ_{SP} are the inverses corresponding to $R_{GV}(q,\delta)$ and $R_{SP}(q,\delta)$; q is regarded as a fixed parameter.

For q = 2 (the binary case) sharper upper bounds are available. The simplest of these is the so called Bassalygo-Elias bound, which we obtain by combining Theorem 6 with the second Johnson bound. Recall that $\frac{w_n}{n} \to v$ and $\frac{c_n}{w_n} \to \kappa$. The Johnson bound shows that $E(v,\kappa) = 0$ for $\kappa \leq v$. Thus by Theorem 6 we have the inequality

$$R(2,\delta) \leq \log 2 - h(\nu),$$

valid for all $\nu \geq \kappa = 1 - \dfrac{\delta}{2\nu}$. By choosing the largest possible ν satisfying this requirement and employing a simple continuity argument we get the following result.

Theorem 16: (Bassalygo-Elias bound)

$$R(2,\delta) \leq \log 2 - h(\tfrac{1}{2} - \tfrac{1}{2}\sqrt{1-2\delta}) \quad 0 \leq \delta \leq \tfrac{1}{2}.$$

This result provides a considerable improvement upon the sphere packing bound, but is still not the best result known. The best upper bound for binary codes know today (1989) is the <u>McEliece-Rodemich-Rumsey-Welch bound</u>, which is obtained from the so called <u>linear programming bound</u>. Unfortunately this result requires a background which we haven't developed here, so we refer the interested reader to the original paper [7] or to Ch. 17 of MacWilliams-Sloane [1].

A recent result by Tsfasman-Vladut-Zink [8] based on an idea of Goppa [9], improves upon the Gilbert-Varshamov bound under certain conditions. The bound reads

$$R(q,\delta) \geq R_{TVZ}(q,\delta) \triangleq (1-\delta - \frac{1}{\sqrt{q-1}}) \log q \quad 0 \leq \delta < 1 - \frac{1}{\sqrt{q-1}}$$

and presupposes that q is an even power of a prime, $q = p^{2s}$. An improvement over the Gilbert-Varshamov bound is obtained in an interval $\delta_1 < \delta < \delta_2$ if $q \geq 49$. The construction uses advanced results from algebraic geometry, and again we refer to the literature [8] - [13] for the details.

Turning to $E(\nu,\kappa)$, we get the following bounds by combining Theorem 4 and Corollary 11 with Lemma 15,

$$E(\nu,\kappa) \leq E_J(\nu,\kappa) = \begin{cases} h(\nu\kappa) - \nu h(\kappa); & 0 \leq \nu \leq \kappa \leq 1 \\ 0; & 0 \leq \kappa < \nu \leq 1 \end{cases}$$

$$E(\nu,\kappa) \geq E_G(\nu,\kappa) \triangleq \begin{cases} h(\nu) - \nu h(\kappa) - (1-\nu)h\left(\dfrac{\nu}{1-\nu}(1-\kappa)\right); & 0 \leq \nu \leq \kappa \leq 1 \\ 0; & 0 \leq \kappa < \nu \leq 1. \end{cases}$$

These bounds are illustrated in fig. 2. None of them is best known. As far as upper bounds are concerned, various improvements have been obtained by Levenshtein [14], [15], Sidelnikov [16] and McEliece-Rodemich-Rumsey-Welch [7]. All of these results are fairly complicated, and will not be derived here. An improvement on the lower bound was recently obtained by Ericson-Zinoviev [18] by using the Tsfasman-Vladut-Zink bound in Theorem 7. The result is

$$E(\nu,\kappa) \geq E_{TVZ}(\nu,\kappa) \triangleq \nu \left[\kappa - \frac{\sqrt{\nu}}{1-\sqrt{\nu}} \right] \log \frac{1}{\nu} \ ; \ \kappa \geq \frac{\sqrt{\nu}}{1-\sqrt{\nu}}$$

where $\nu = q^{-1} = p^{-2s}$; p prime, s integer. This new bound improves the bound $E_G(\nu,\kappa)$ in a certain range $\kappa_1 < \kappa < \kappa_2$ provided $\nu = p^{-2s} \leq \frac{1}{81}$.

References

[1]   F.J. MacWilliams and N.J.A. Sloane, "The theory of error-correcting codes", North Holland, 1977

[2]   J.H. van Lint, "Introduction to coding theory", Graduate texts in mathematics, vol. 86, New York: Springer Verlag, 1982

[3]   Th. Beth, D. Jungnickel and H. Lenz, "Design theory", Wissenschaftsverlag, Mannheim, Wien, Zürich, 1985

[4]   E.M. Gilbert, "A comparison of signalling alphabets", Bell syst. techn. journal, vol. 31, pp. 504-522, 1952

[5]   R.R. Varshamov, "Estimate of the number of signals in error-correcting codes",. Dokl. Akad. Nank SSSR, vol. 117, pp. 739-741, 1957

[6]   E.L. Blokh and V.V. Zyablov, "Linear concatenated codes", Moscow 1982, (in Russian)

[7]   R.J. McEliece, E.R. Rodemich, H.C. Rumsey, Jr. and L.R. Welch, "New upper bounds on the rate of a code via the Delsarte-MacWilliams inequalities", IEEE Trans. Info. Theory, vol. IT-23, pp. 157-166, 1977

[8]   M.A. Tsfasman, S.G. Vladut and Th. Zink, "Modular curves, Shimura curves and Goppa codes, better than Varshamov-Gilbert bound", Math. Nachr., vol. 104, pp. 13-28, 1982

[9]   V.D. Goppa, "Codes on algebraic curves", Soviet Math. Doklady, vol. 24, pp. 170-172, 1981

[10]   V.D. Goppa, "Algebraic geometry codes", Math. USSR Isvestiga, vol. 21, pp. 75-91, 1983

[11]   G.L. Katsman, M.A. Tsfasman and S.G. Vladut, "Modular curves and codes with a polynomial construction", IEEE Trans. Info. Theory, vol. IT-30, pp. 353-355, 1984

[12]   M.A. Tsfasman, "Goppa codes that are better than the Varshamov-Gilbert bound", Probl. of Info. Transm., vol. 18, pp. 163-165, 1982

[13]   J.H. van Lint and T.A. Springer, "Generalized Reed-Solomon codes from algebraic geometry", IEEE Trans. Info. Theory, vol. IT-33, pp. 305-309, May 1987

[14]   V.I. Levenshtein, "Upper bound estimates for fixed-weight codes", Probl. of Info. Transm., vol. 7, pp. 281-287, (Engl. transl.), 1971

[15]   V.I. Levenshtein, "Minimum redundancy of binary error-correcting codes", Probl. of Info. Transm., vol. 10, pp. 110-123, (Engl. transl.), 1974

[16]   V.M. Sidelnikov, "Upper bounds for the number of points of a binary code with a specified code distance", Probl. of Info. Transm., vol. 10, pp. 124-131 (Engl. transl.), 1974

[17]   L.A. Bassalygo, "New upper bounds for error-correcting codes", Probl. of Info. Transm., vol. 1, pp. 32-35 (Engl. transl.), 1965

[18]   T. Ericson and V.V. Zinoviev, "An improvement of the Gilbert bound for constant weight codes", IEEE Trans. Info. Theory, vol. IT-33, pp. 720-723, September 1987

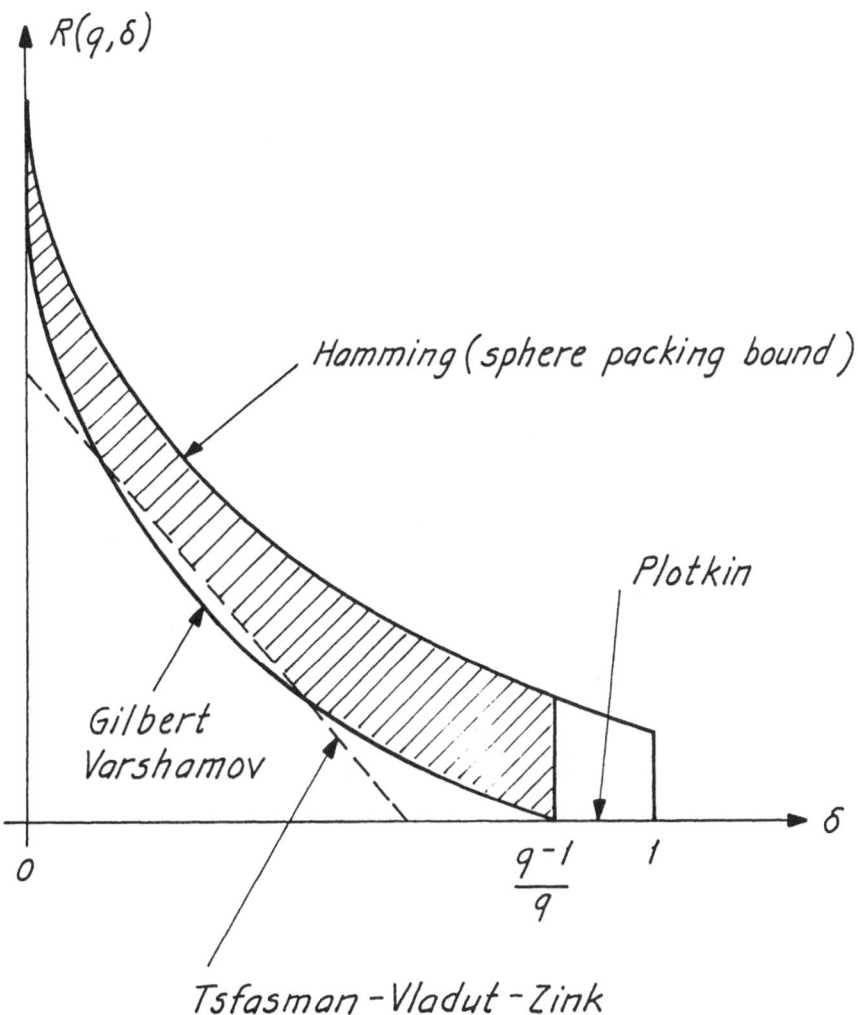

Figure 1
Exponential bounds on the size of a code; q ≥ 49. The true exponent
R(q,δ) is somewhere in the dashed region.

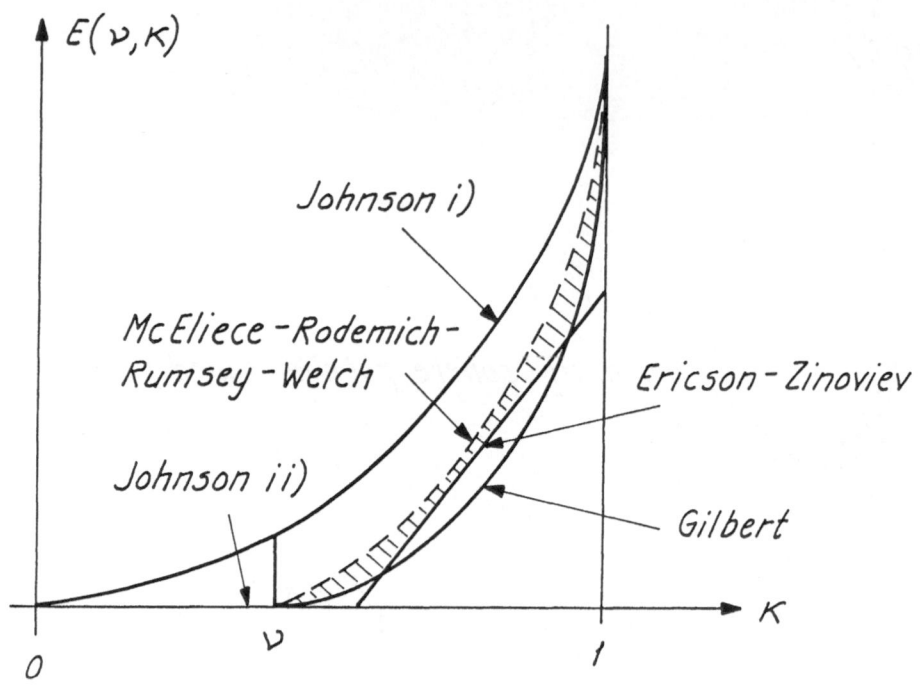

Figure 2
Exponential bounds for constant weight codes. The true exponent
$E(v,\kappa)$ is somewhere in the dashed region.

Group Codes for the Gaussian Channel

Ingemar Ingemarsson
Department of Electrical Engineering
Linköping University, S-581 83 Linköping, SWEDEN

ABSTRACT

This paper gives a tutorial overview over the results from the research on Group Codes for the Gaussian Channel. Such a code represents a finite set of signals. It may thus be regarded as combined coding and modulation.The channel is described by the vector model. A Group Code for the Gaussian Channel is a set of vectors with complete symmetry. The symmetry is required to give the same probability of error (with a maximum-likelihood receiver) for every code vector.

The set of vectors is obtained by operating with a multiplicative group of matrices on a given initial vector. The performance of the set is measured by regarding the probability of error (with a maximum-likelihood receiver) and the smallest mutual Euclidean distance between the vectors. Some general results regarding these measures are given.

The Group Codes for the Gaussian Channel behave differently depending on whether the group is commutative or not. Commutative Group Codes are closely related to linear algebraic codes combined with phase modulation. Non-commutative Group Codes have potential for better performance but are less well understood. The largest known class is Permutation Modulation.

Application of Group Codes for specific channels models are discussed.

1. INTRODUCTION

1.1 The physical channel and a mathematical model

In the center of a communication system there is always a physical communication channel. This channel is described by a (mostly analog) channel model involving the relation between the input and output signals to and from the channel. Most physical channels admit signals which are both time-continuous and have continuous amplitude distribution. When the channel is used for digital communication (i.e. with a finite number of possible messages per time unit) the messages ought to be represented by a finite set of signals in a finite time interval. One important class of such representations is <u>Group Codes for the Gaussian Channel</u>.

The channel model to be used here, the Gaussian Channel, assumes additive white noise which is independent of the signal. The channel is conveniently described by the vector model; see Wozencraft & Jacobs [5], particularly chapter 4.

1.2 Coding and modulation

In many cases the representation of the set of messages by a set of signals is done in two steps. First the messages are represented by a set of digital code words, i.e. vectors over a finite set. The components of these vectors are then represented by signals. This last representation is the same for all of the vector components. The first representation may be called coding, the second modulation. Historically there has been a tremendous amount of research done in coding theory and coding techniques and comparatively little in modulation. The two steps have also been considered separately. Recently, however, Ungerboeck and others have designed schemes for combined coding and modulation.

Group Codes for the Gaussian Channel may be regarded as combined coding and modulation. Instead of the two-step procedure described above the set of messages are directly represented by a set of signals to be transmitted over the channel. The class of representations is in fact general enough to include most combinations of digital codes and subsequent modulation.

1.3 ML detection and error probability

The messages to be transmitted are assumed to have equal probability. The detector is a maximum-likelihood (ML) detector, which yields the lowest possible error probability for this case. In the vector space the ML-detector can be described as a minimum-distance detector: the detector chooses the signal vector with the smallest Euclidean distance to the received vector. This detector thus divides the vector space into decision regions, one for each possible transmitted vector. The probability of a correct decision is equal to the probability that the received vector falls within the same decision region as the transmitted vector. Intuitively it is desirable that these probabilities are equal. If this shall be the case for all signal-to-noise ratios the decision regions ought to be congruent.

The assumption of congruent decision regions is crucial to Group Codes for the Gaussian Channel. Signal sets with congruent decision regions are called completely symmetric by Wozencraft & Jacobs, [5], p. 263. Slepian [1], p. calls them equipunctual configurations. The names indicate that all code vectors (or code words) are on equal footing, they have the same error probability and the same set of distances to the other code vectors. If the set of vectors is finite they must all have the same length, i.e. the corresponding signals have the same energy. The code vectors terminate on an n-dimensional sphere.

Shannon [2] and Slepian [3] calculated lower bounds on the error probability for code vectors on a sphere by assuming that the decision regions are circular cones. These bounds are used as reference to judge the quality of codes. The actual (or approximate) error probability is compared to the bound for the same signal-to-noise ratio, dimensionality and number of code vectors. Another way to make the comparison is to calculate the increase in signal-to-noise ratio needed for the code to yield the same error probability as the lower bound. This increase is in the order of 2 db or less for good codes.

1.4 Outline of paper

In Section 2 below we will define Group Codes for the Gaussian Channel as a set of vectors obtained by a group of orthogonal matrices operating on an initial vector. Some general results from matrix algebra will be useful in characterizing the codes. The performance of the codes is measured in terms of the error probability or more coarsely in terms of the mutual distances between the code vectors. Some general results regarding the set of mutual distances and bounds on the error probability will be given. The initial vector problem, i.e. the choice of the initial vector which minimizes the error probability or maximizes the minimum mutual distance between the code vectors, will be discussed. We will also give some results regarding the existence of Group Codes for particular numbers of code vectors and dimensions.

The Group Codes behave differently depending on whether the group is commutative or not. In Section 3 we will discuss Commutative Group Codes. These will be shown to be closely related to linear algebraic codes combined with antipodal or phase modulation. In fact Zetterberg [8] introduced what he called polyphase codes before Slepian published his results on Group Codes for the Gaussian Channel [12]. Section 3 also includes an interesting generalization of Group Codes to include infinitely many code words. These codes are called Continuous Group Codes and tie coding to frequency modulation and other schemes for the transmission of analog information.

Non-commutative Group Codes, which will be treated in Section 4, have potential for better performance than Commutative Group Codes but are less well understood. The largest known class is Permutation Modulation, which is a class of Group Codes for the Gaussian Channel, despite the name. In a way Permutation Modulation resembles Commutative Group Codes representing non-redundant (i.e. non-coded) algebraic codes. The maximum-likelihood detector is easy to describe and implement. Algebraic coding, i.e. restriction to a sub-group, may be used in both cases. For Permutation Modulation, though, the use of subgroups remains to be investigated. Besides Permutation Modulation little is done on non-commutative Group Codes. Ottoson, [13] and [15], has treated codes based on products of permutation groups and commutative groups. Zetterberg and Brändström [28] have published results for Group Codes based on quarternions.

Section 5 covers application of Group Codes to specific channels. The obvious application of the codes to single-user Gaussian channels will not be discussed. Other applications include fading channels [16], discrete channels [23], broadcast channels [30], [34], [36] and jamming channels [38].

One important use of Group Codes which will not be discussed here is source coding. Permutation Modulation, in particular, have been shown to offer optimum coding for a large class of sources.

2. GENERAL GROUP CODES FOR THE GAUSSIAN CHANNEL

2.1 The vector model of the additive Gaussian channel

The M signals transmitted on the Gaussian channel are represented by vectors X_1, ... ,X_M in an n-dimensional Euclidean vector space. (See Wozencraft & Jacobs [5] for a thorough treatment of the vector model for the Gaussian channel.) The vectors are assumed to have equal *a priori* probability. The transmitted signals are disturbed by additive white noise with bilateral spectral density N/2. The noise is modelled as a noise vector Z, the components of which are independent Gaussian random variables with zero mean and variance N/2. The received vector Y is the sum of the transmitted vector X and the noise vector.

$$Y = X + Z$$

The maximum-likelihood detector chooses the code vector closest (at shortest Euclidean distance) to the received vector. The set of received vectors thus assigned to the transmitted vector X_i thus form a decision region R_i.

2.2 Definitions

The code (i.e. the set of possible transmitted vectors) is invariant under multiplication with a group G of orthogonal matrices. The code is defined as a Group Code for the Gaussian Channel (Slepian [1] and [12]) if there are matrices O_1, ... ,O_M in G such that the code can be formed by the multiplication of any particular code vector with the matrices O_1, .. , O_M. Note that in general the matrices O_1, ... ,O_M do not form a group.

The rate of the code is defined as:

$$R = \frac{\log_2 M}{n} \qquad \text{bits/dimension}$$

The decision regions for the vectors in a Group Code are all congruent and all vectors have the same length.

$$nS = |\underline{X}_i|^2 \qquad (1)$$

Here S is the average _signal power_, i.e. the energy per dimension.

The code may be obtained by multiplying any one of the code vectors, say $\underline{X}_1$, with the group G. The vector $\underline{X}_1$ is called the _initial vector_.

$$\{\underline{X}_i\} = G\underline{X}_1 \qquad (2)$$

By using the set of matrices O_i defined above we obtain the following relation.

$$\underline{X}_i = O_i\underline{X}_1 \qquad (3)$$

The set of Euclidean distances between the code vectors is the same as the set of distances from any one of the code vectors to the other code vectors. We will use the following notation for the distances in this set.

$$d_i = |\underline{X}_i - \underline{X}_1| \qquad (4)$$

From this we easily obtain the following relation between d_i and the _scalar product_ ρ_i between $\underline{X}_i$ and $\underline{X}$.

$$\rho_i = \underline{X}_i^T\underline{X}_1 \qquad (5)$$

$$d_i^2 = 2nS - 2\rho_i \qquad (6)$$

The number of code vectors, M, is usually less than the cardinality of G. The initial vector is invariant under multiplication with the subgroup $\mathcal{H}$ of G.

$$\underline{X}_1 = \mathcal{H}\underline{X}_1 \qquad (7)$$

It is then easy to show [19] that the matrices $O_1, \dots ,O_M$ can be chosen as coset leaders for the cosets to $\mathcal{H}$ in G. Thus the number of code words is the ratio between the cardinalities of G and $\mathcal{H}$.

$$M = ||G||/||\mathcal{H}|| \qquad (8)$$

If $\mathcal{H}$ is a normal (self-conjugate) subgroup of $\mathcal{G}$ then it reduces to the unit matrix (if the code spans the n-space) and the set $\{O_i\}$ is identical with $\mathcal{G}$. This is seen by using (3) and (7):

$$X_i = O_i X_1 = O_i \mathcal{H} X_1 = \mathcal{H} O_i X_1 = \mathcal{H} X_i$$

Thus every code vector is left invariant by the group $\mathcal{H}$ and since the code spans the n-dimensional space the whole space is left invariant by $\mathcal{H}$. Then $\mathcal{H}$ can only contain the unit matrix.

This is for example the case when $\mathcal{G}$ is commutative, which is discussed in Section 3 below.

2.3 Real-reducibility

The error probability of the code is unchanged if the code is multiplied with an orthogonal matrix A, which only rotates the code.

$$X' = A\mathcal{G}X_1 = A\mathcal{G}A^{-1}X'_1$$

The codes $\{X\}$ and $\{X'\}$ are called equivalent. If the group $\mathcal{G}$ is real-reducible [12] then there is a matrix A such that every matrix in the group $A\mathcal{G}A^{-1}$ is pseudodiagonal:

$$A\mathcal{G}A^{-1} = \begin{pmatrix} \mathcal{G}_1 & Z_1 \\ Z_2 & \mathcal{G}_2 \end{pmatrix}$$

Here $\mathcal{G}_1$ and $\mathcal{G}_2$ are groups of orthogonal matrices of smaller dimension than $\mathcal{G}$. Z_1 and Z_2 are matrices with zero components only. If $\mathcal{G}$ is real-reducible then the Group Code is equivalent to a direct sum of Group Codes with lower dimensionalities. This means that the initial vector can be divided into parts and that the matrices generating the code operate on these parts separately. We will call these codes Real-reducible Group Codes.

If $\mathcal{G}_1$ or $\mathcal{G}_2$ consist of the unit matrix alone then the corresponding part of the initial vector is left invariant by $\mathcal{G}$. The Group Code is then called a Planar Code. This is usually avoided. In most cases we will assume that the codes are Non-Planar Group Codes, i e they are not equivalent to a direct sum of group codes one or more of which is a fixed vector.

2.4 Performance of General Group Codes

2.4.1 Error probability and minimum distance

Since the decision regions for a Group Code are congruent the error probability is obtained by the following relation.

$$P_e = 1 - \int_{R_1} \frac{1}{\sqrt{\pi N n}} e^{-\frac{|Z - X_1|^2}{N}} dZ \tag{9}$$

Here R_1 is the decision region that belongs to the initial vector X_1. The decision region is bounded by hyperplanes bisecting and perpendicular to the difference vector between X_1 and the neighboring code vectors. Hence we obtain the following relation.

$$P_e \geq \int_{d_{min}/2}^{\infty} \frac{1}{\sqrt{\pi N}} e^{-\frac{z^2}{N}} dz \tag{10}$$

Here d_{min} is the minimum Euclidean distance between any two code vectors.

$$d_{min} = \min_{i=[2,M]} d_i \tag{11}$$

It is apparent from the relation above that d_{min} is an approximate measure of the performance of the code. The error probability is of course the performance measure we would really want to use, but numerical difficulties often make it necessary to resort to d_{min}.

2.4.2 The initial vector problem

With these two performance measures we get two optimization problems. In order to find the best Group Code we start with a given group G of orthogonal matrices and try to find the initial vector yielding the highest d_{min} or, preferably, the lowest error probability with given signal power S and give dimensionality n. Both of these optimization problems have been attacked by several researchers [4], [12], [14], [18], [20], [25], [26], [31], [33], [39]. Still both problems are unsolved for general Group Codes. As we will see below, though, there has been substantial progress for particular classes of Group Codes.

2.4.3 The union bound

In addition to the lower bound on the error probability based on d_{min} which we have seen above there is also an upper bound based on the union bound:

$$P_e < P_u = \sum_{\underline{X}_i \neq \underline{X}_1} \int_{d_i/2}^{\infty} \frac{1}{\sqrt{\pi N}} \, e^{-\frac{z^2}{N}} \, dz \qquad (12)$$

If we regard $\underline{X}_i$ as a random vector which with equal probability is equal to any one of the code vectors, then ρ_i in (5) is a random variable. The distribution of this random variable is almost Gaussian for large codes [14]. The mean is zero for a non-planar code. The union bound P_u can be expressed as a weighted sum of the moments of the random variable ρ_i (ref. [14] and [21]). The coefficients in the sum are all positive and thus P_u increases with increasing moments.

2.4.4 The Configuration Matrix

The configuration matrix of a group code is the M by M matrix C whose elements are the normalized scalar products between $\underline{X}_i$ and $\underline{X}_j$. [12] Thus

$$C_{ij} = (nS)^{-1} (\underline{X}_i, \underline{X}_j) \qquad (13)$$

Each row (or column) in the configuration matrix is a permutation of each other row (or column) in the matrix. Further C is real, symmetric non-negative definite and of rank n. The diagonal elements are unity and all elements are of magnitude no greater than one. Ian Blake has shown (ref.[22] and [24]) that the configuration matrix of a code generated by a real-irreducible matrix group is a scalar multiple of an idempotent matrix. The configuration matrix seems to be related to the problem of finding the initial vector that maximizes the minimum Euclidean distance between the code vectors, but so far no results in this direction have been published.

2.4.5 Asymptotic performance for real-reducible Group Codes

In an unpublished report [7] Peterson and Kasami has determined bounds on the error exponent for codes which are equivalent to a direct sum of codes on lower-dimensional spheres. They have considered direct sums of 1-, 2-, 3- and 5-dimensional codes and compared the results to codes which are not equivalent to direct sums of lower-dimensional codes. Even though Peterson and Kasami do not mention the term Group Codes (they use "Polyphase Codes"), their results are certainly applicable as bounds on the asymptotic performance of Real-Reducible Group Codes.

The classes of codes considered in [7] are:

Class S: All code vectors terminate on an n-dimensional sphere or inside the sphere. (The bounds are equal for these two cases.)

Class D2: Antipodal binary codes. Each component of a code vector is either $\sqrt{S}$ or $-\sqrt{S}$.

Class Dq: Polyphase codes. Each pair of components of a code vector is located at one of q equidistant points on a circle with radius $\sqrt{2S}$.

Class Cm: The codes are direct sums of m-dimensional spherical codes. Each of the sphere has radius $\sqrt{mS}$.

The error exponent E is defined as follows.

$$E = \lim_{n \to \infty} \left(-\frac{1}{n} \ln P_e\right) \qquad (14)$$

Peterson and Kasami use rates R measured in nats/dimension, thus:

$$R_n = \lim_{n \to \infty} \left(-\frac{1}{n} \ln M\right) \qquad \text{nats/dimension}$$

The larger E, the better the codes. Peterson and Kasami give bounds on E which then are bounds on the error exponent for Real-Reducible Group Codes. The different bounds are:

1. The Sphere Packing bound.
2. The Plotkin Bound.
3. The Random-Coding Bound.
4. The Expurgated Random-Coding Bound.

The randomization for bounds 3 and 4 is done over all possible codes in each class, not only over Group Codes. It remains to be proved that these lower bounds are also valid when the codes are restricted to Group Codes for the Gaussian channel. The upper bounds 1 and 2 are of course upper bounds also for the error exponent for Group Codes.

The results are shown in Figures 1 to 7 for different values of the signal-to-noise ratio:

$$A = \sqrt{2S/N}$$

E is shown as a function of the rate, R_n, of the code. The letters U, B and Q denotes the intersections between the different bounds. C is the capacity. See Figure 1. The curves are labeled by the number which indicates the type of bound, followed by the code class. Points are labeled by the point name, followed by the class of codes.

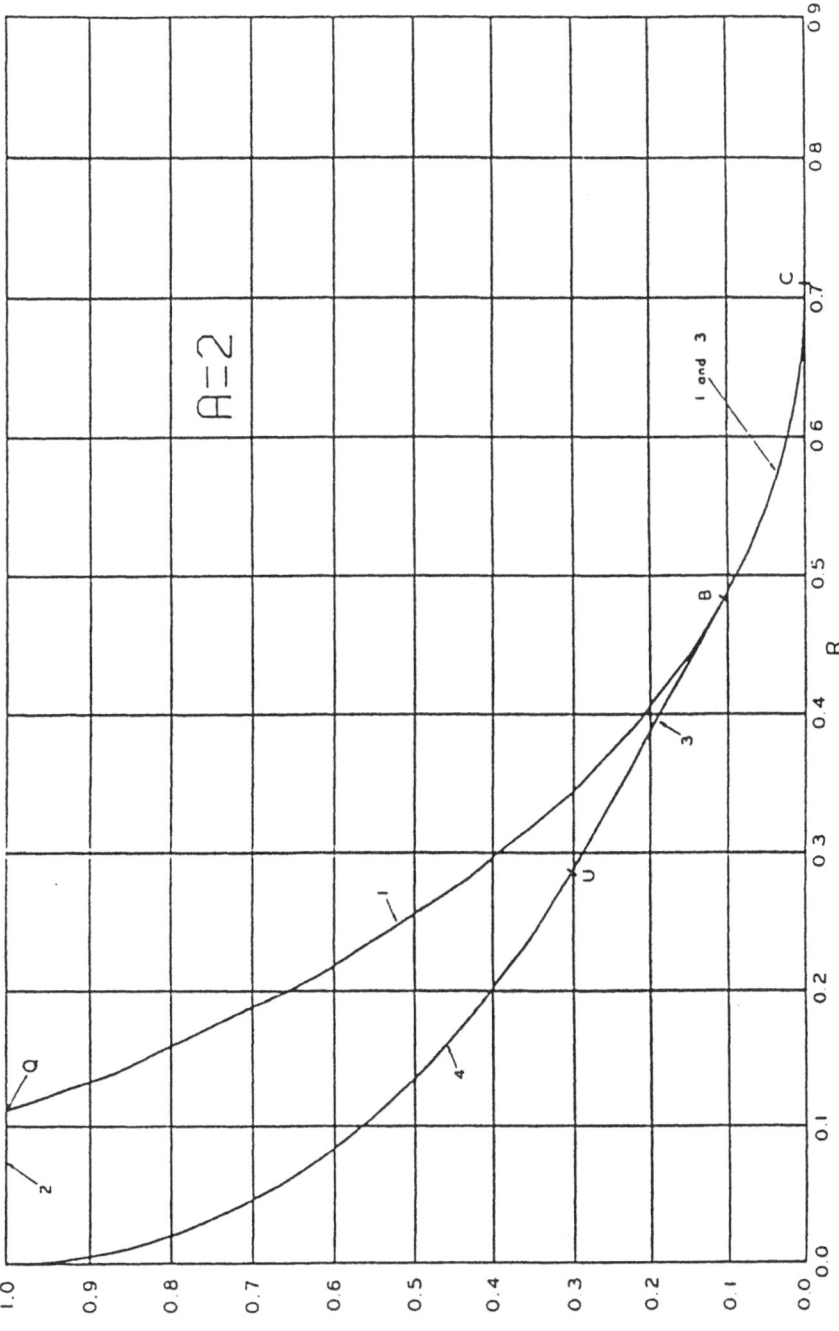

Figure 1 A typical set of reliability bounds (case C-2 for A = 2)

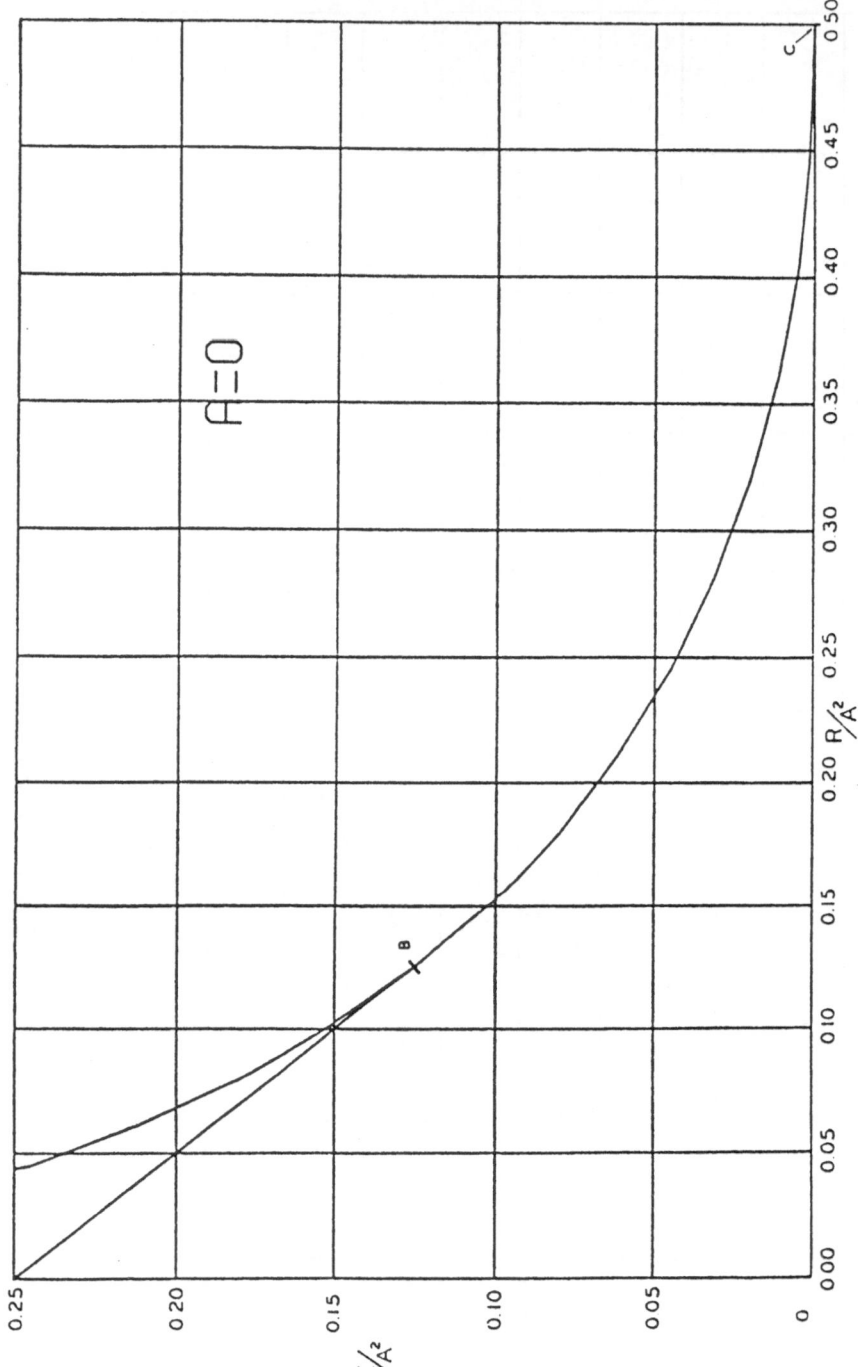

Figure 2 Asymptotic reliability for small A

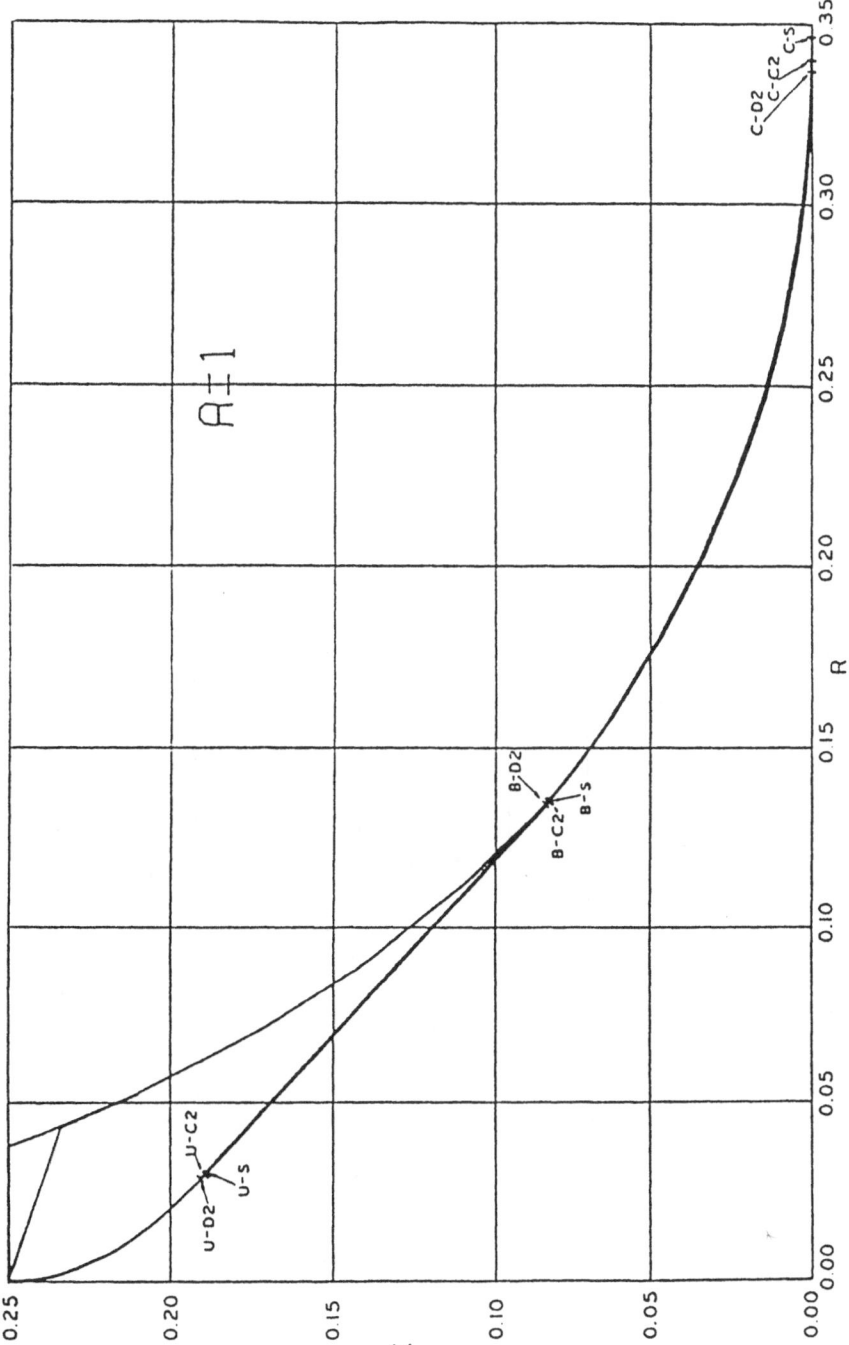

Figure 3 Reliability bounds for A = 1

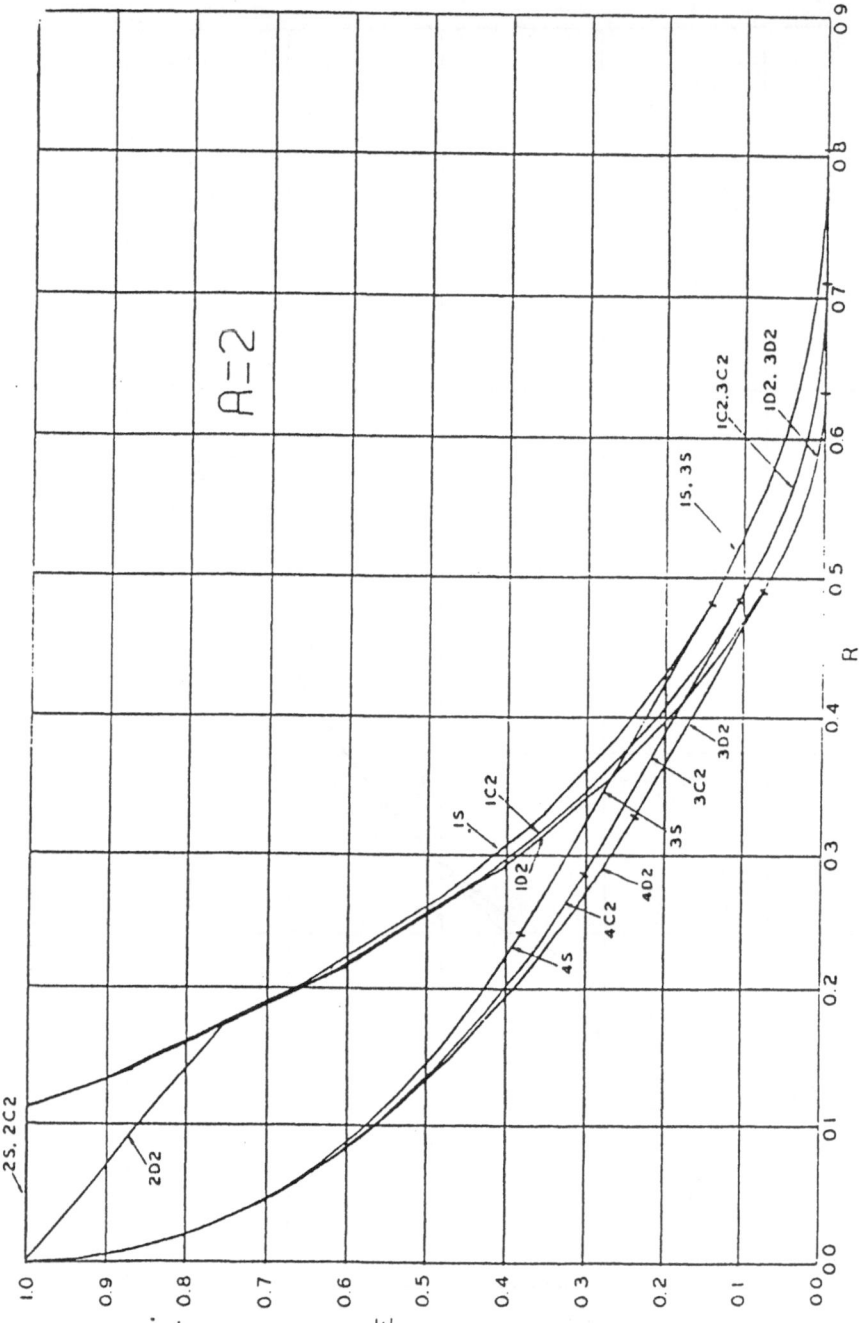

Figure 4 Reliability bounds for A =2

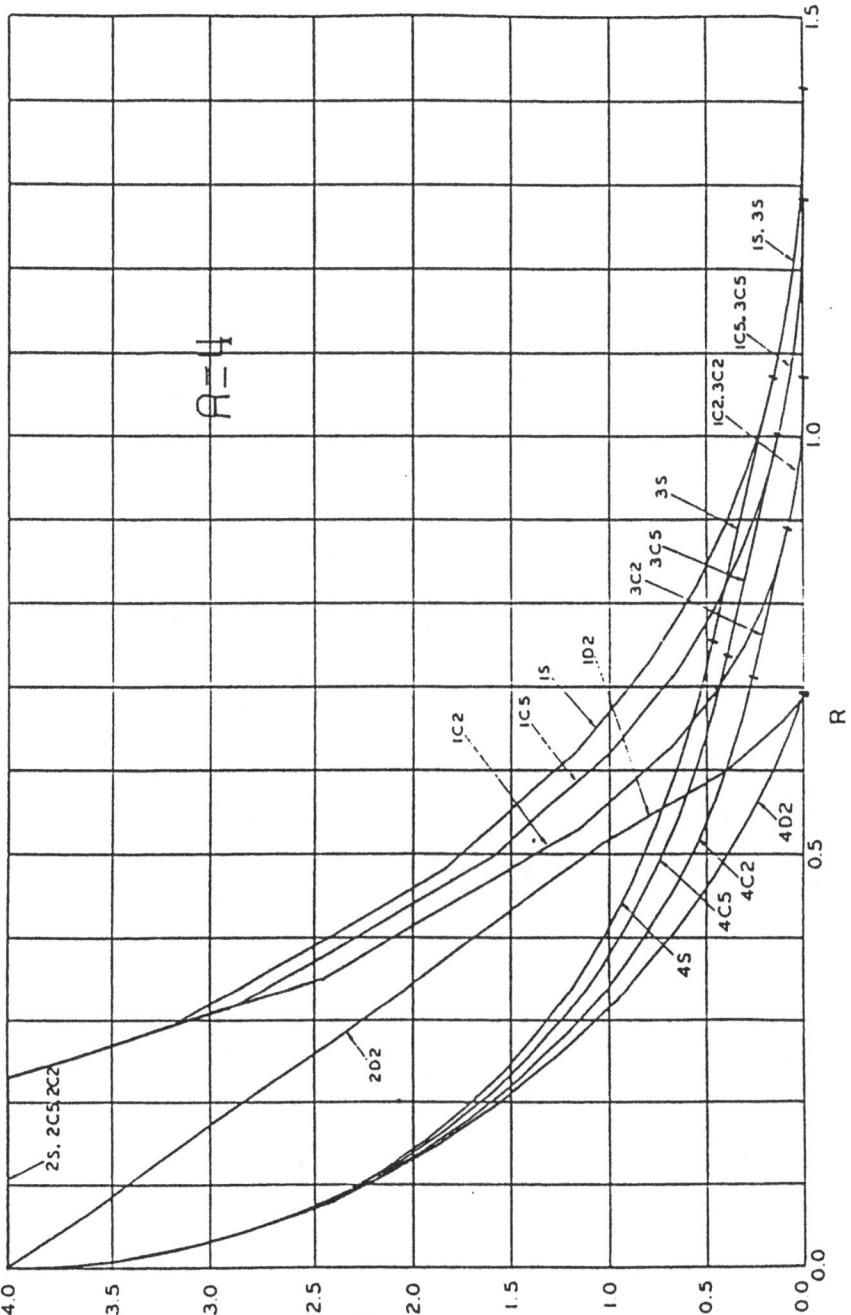

Figure 5 Reliability bounds for A = 4

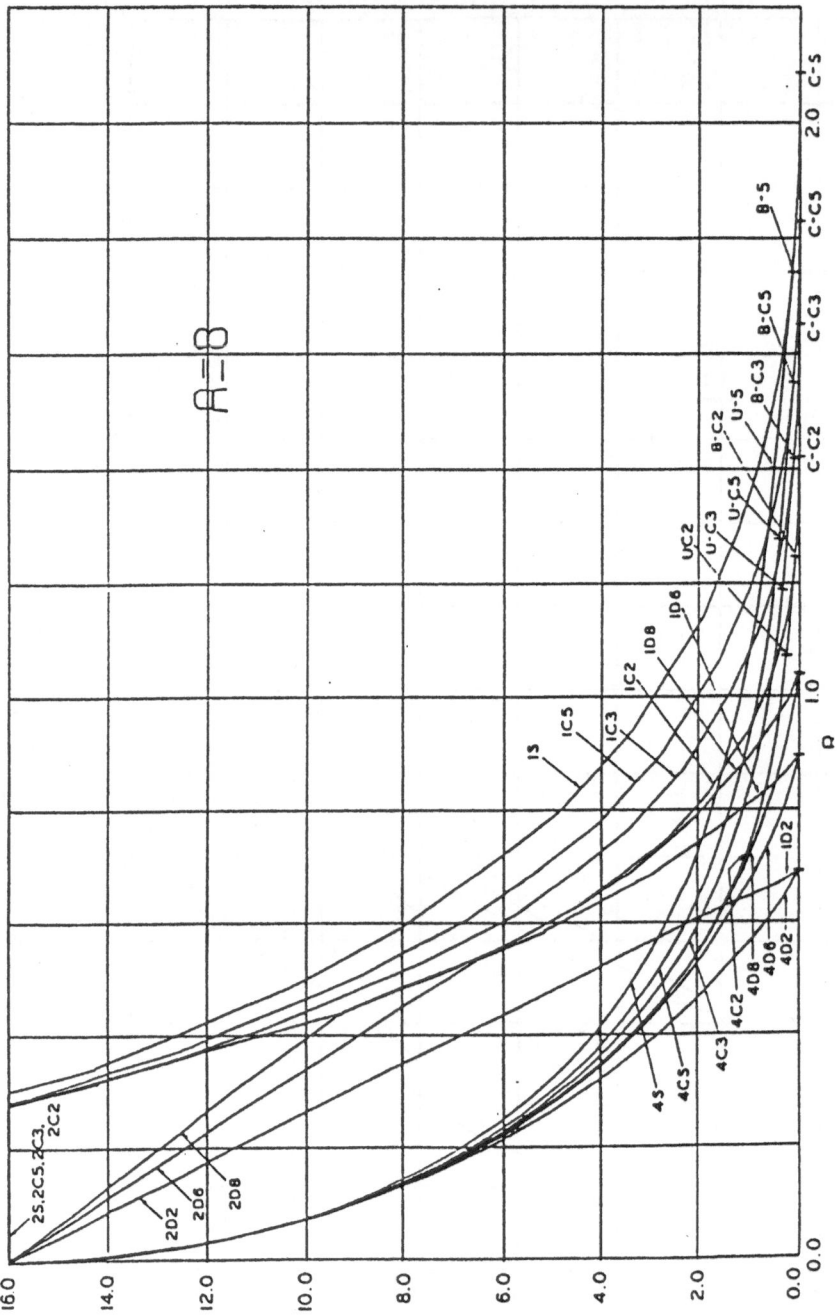

Figure 6 Reliability bounds for A = 8

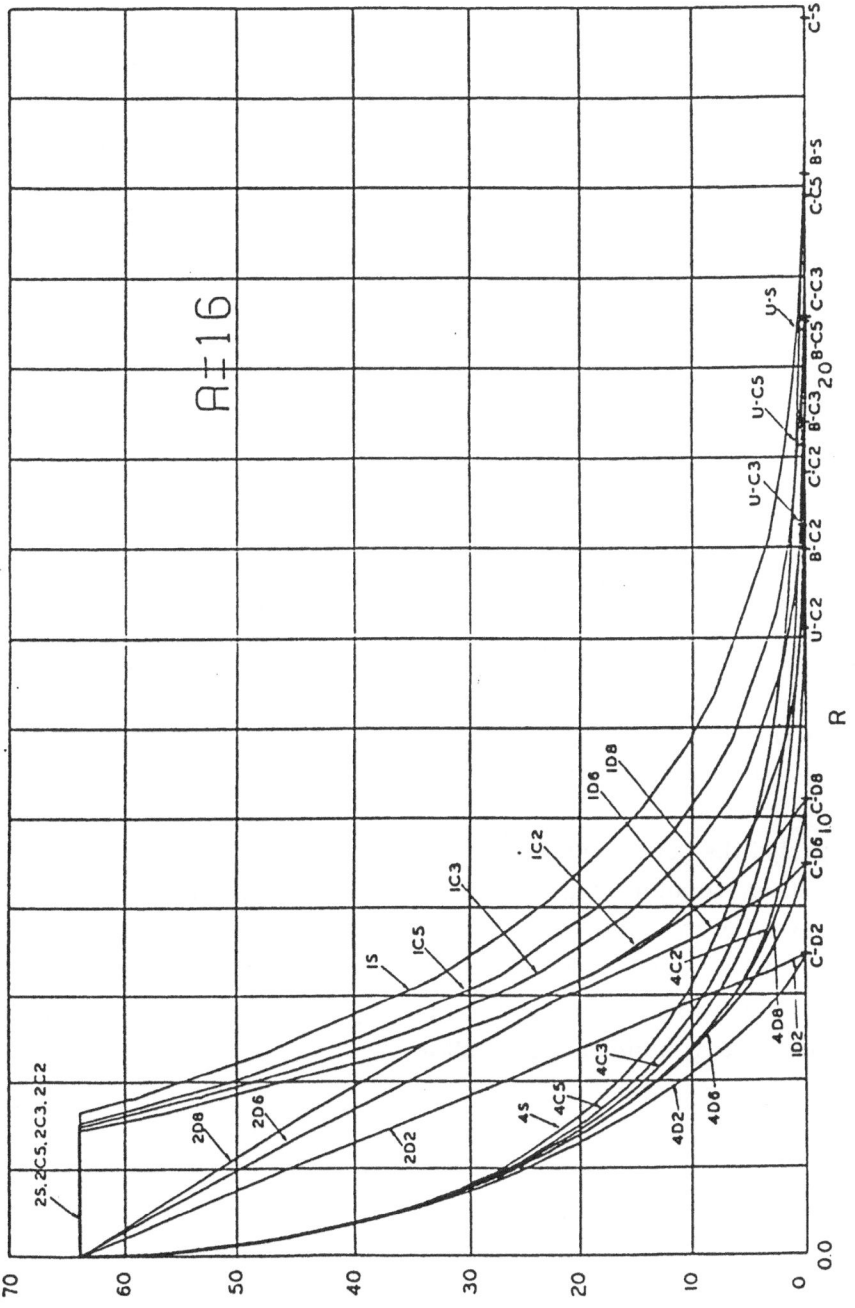

Figure 7 Reliability bounds for A = 16

The point where E reaches zero is the capacity for codes which are direct sums of random codes on lower-dimensional spheres. This capacity is shown in Figure 8 as a function of the signal-to-noise ratio. That this is also the capacity for Real-Reducible Group Codes remains to be proven.

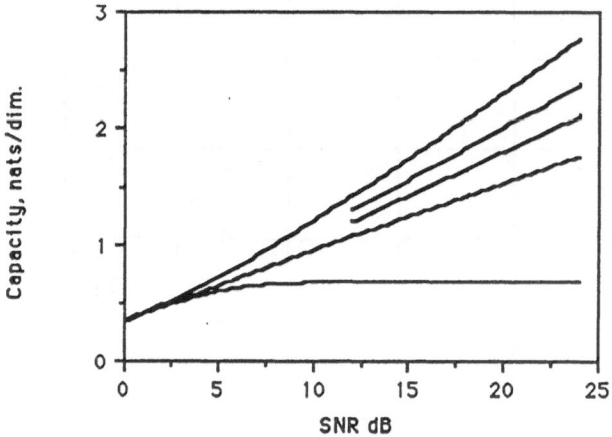

Figure 8 Capacity for codes of classes D2, C2, C3, C5 and S respectively (from lowest to highest)

As is expected the performance is hurt by the reducibility of the codes. It is interesting, though, that the increase in performance when the dimensionality of the subcodes in the direct sum goes from, say 2 to 5, is greater than the increase from 5-dimensional to real-irreducible subcodes. Thus it may be worthwhile studying Group Codes which are direct sums of, say, 5-dimensional subcodes.

2.5 Existence of General Group Codes

Group Codes do not exist for all combinations of n and M. We have seen before that M must be a factor in the cardinality of G. Thus when the code is generated by a group of odd prime cardinality, M must be equal to this number. Moreover all such groups are commutative and thus the corresponding codes are equivalent to a direct sum of 2-dimensional codes (see Section 3 below). Hence the dimensionality is even.

Using similar reasoning we obtain the table below.

n	M	Note	Reference
even	any odd prime >n	G is cyclic	[14],[19],[21]
even	any nonprime >n		[14],[19],[21]
3	any even number >n	The initial vector yielding max. d_{min} is known	[31], [33]
odd>3	some odd nonprime >n		[19], [29]

Table 1. Existence of non-planar Group Codes.

3. COMMUTATIVE GROUP CODES

3.1 Definition

A Group Code for the Gaussian Channel is called a <u>Commutative Group Code</u> if the group G generating the code is commutative. Then the group $\mathcal{H}$ leaving the initial vector invariant consists of the unit matrix only and the number of codewords, M, is equal to the cardinality of G. The set $\{O_i\}$ is identical with G.

3.2 Real-reducibility

Commutative matrix groups are always real-reducible. In fact every Commutative Group Code is equivalent to a direct sum of 1- and 2-dimensional group codes ([21], Theorem 2.1). Thus the matrices O_i may be written in a pseudodiagonal form:

$$O_i = \left[\begin{pmatrix} \cos q_{i1} & \sin q_{i1} \\ -\sin q_{i1} & \cos q_{i1} \end{pmatrix} \cdots \begin{pmatrix} \cos q_{ir} & \sin q_{ir} \\ -\sin q_{ir} & \cos q_{ir} \end{pmatrix} \pm 1 \cdots \pm 1 \right] \tag{15}$$

The square brackets denote a pseudodiagonal matrix in which the diagonal elements are r 2 by 2 orthogonal matrices and (n-2r) ± 1.

3.3 The additive group of rotation vectors

The effect of O_i can thus be described by a <u>rotation vector</u>:

$$\mathbf{q}\,i = (q_{i1}, \ldots, q_{ir}, q_{i\,r+1}, \ldots) \tag{16}$$

Here $q_{i\,r+1}, \ldots$ are 0 or π.

The action of O_i on the initial vector $\underline{X}$ can be depicted as in Figure 9.

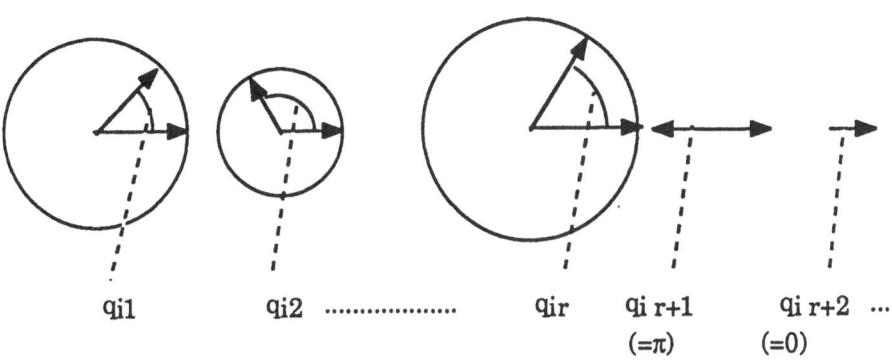

qi1   qi2 ....   qir   qi r+1   qi r+2 ...

                                                  (=π)   (=0)

Figure 9. The projection of two vectors of a Commutative Group Code

The rotation vector of the product of two matrices is the sum (mod 2π) of their rotation vectors. Thus we obtain an additive group of rotation vectors, simply isomorphic with G. Addition is defined as component-wise addition mod 2π.

A commutative multiplicative group is equal to the direct product of cyclic subgroups of prime power orders ([21], p.216). The group of rotation vectors is thus equal to the sum (mod 2π) of cyclic subgroups of prime power orders. Each subgroup is generated by a vector λ. The rotation vector for a particular matrix O may then be expressed as in (17).

$$\underline{q} = \sum_{j=1}^{k} m_j \lambda_j \ \text{mod } 2p = \underline{m} \Lambda \text{ mod } 2p \qquad (17)$$

Here Λ is a matrix called the <u>rotation generator matrix</u>. It has the row vectors $\underline{\lambda}_1, ..., \underline{\lambda}_k$. The rotation vector λ_j generates a cyclic group of prime power order q_j. Thus the components of λ_j have to be multiples of $2\pi/q_j$.

$\underline{m}$ is the integer row vector:

$$\underline{m} = (m_1, ..., m_k) \qquad (18)$$

Here: $\qquad 0 < m_i < q_i$

The number of code vectors is:

$$M = \prod_{i=1}^{k} q_i \qquad (19)$$

We have thus arrived at a description of Commutative Group Codes which closely resembles linear codes. The additive group (mod 2π) of rotation vectors $\{q_i\}$ corresponds to the set of code words. The vector $\underline{m}$ contains the information symbols and the matrix L corresponds to the generator matrix.

3.4 Commutative Group Codes representing linear algebraic codes

If all the periods q_i are equal to the same prime number p then we can define the integer vectors $\underline{C}$ as follows.

$$\underline{C} = \frac{p}{2\pi} \, \underline{q} \qquad (20)$$

Then multiplication of matrices and addition of rotation vectors mod 2π corresponds to addition mod p of the associated vectors $\underline{C}$. In this way we have obtained a Commutative Group Code which naturally represents a linear algebraic code over GF(p).

3.5 Continuous Group Codes

Commutative Group Codes can be generalized to Continuous Group Codes for the Gaussian Channel simply by replacing the integers m_i in (17) with real numbers in the interval [0, q_i). The vector $\underline{X}$ is then a continuous function of the k real variables m_1, ..., m_k. This describes a simply connected signal locus on the n-dimensional sphere. (c f [5], Chapter 8). Note that the signal locus is described by the same rotation generator matrix Λ as before. Thus the locus interpolates the corresponding finite code, i e the set of points in the n-dimensional space.

The signal locus has the same equipunctual property as Group Codes in general: every point on the locus has the same neighborhood. In two dimensions there is one such line: the circle. The dimensionality must be even (c f Figure 9). In four dimensions there are infinitely many signal loci with the equipunctual property, each corresponding to a different 1 by 2 matrix Λ.

Continuous Group Codes may be used to communicate a set of continuously distributed variables $\{m_j\}$. The receiver may still contain a maximum-likelihood detector. This means that the estimation of the transmitted vector is the point on the signal locus closest to the received vector. This corresponds to a set of estimates of the variables $\{m_j\}$. The performance criterion in this case may be the mean-square error.

Continuous Group Codes for the Gaussian Channel are further discussed in [14], Chapter 5.

4. NON-COMMUTATIVE GROUP CODES

4.1 Permutation Modulation

4.1.1 Definition

The first published Group Code for the Gaussian Channel, <u>Permutation Modulation</u> [4], is generated by a non-commutative group of orthogonal matrices.

The code vectors in Permutation Modulation, Variant I, are all distinct vectors obtained by permutation of the components of the initial vector. The group G in this case is the matrix representation of the symmetric group, i e the group of all permutations on n letters. The group $\mathcal{H}$ is the matrix representation of the group of all permutations which leaves the initial vector invariant.

The initial vector $\underline{X}_1$ has the form:

$$X_1 = (\mu_1, ...,\mu_1, \mu_2, ...,\mu_2, ...,\mu_k, ..., \mu_k) \quad (21)$$
$$\leftarrow m_1 \rightarrow \quad \leftarrow m_2 \rightarrow \quad .... \quad \leftarrow m_k \rightarrow$$

The cardinality of G is n! and the cardinality of $\mathcal{H}$ is :

$$||\mathcal{H}|| = \prod_{i=1}^{k} m_i!$$

where:

$$n = \sum_{i=1}^{k} m_i$$

Thus by (8):

$$M = \frac{n!}{\prod\limits_{i=1}^{k} m_i!} \quad (22)$$

In Permutation Modulation, Variant II, the code consists of all distinct vectors formed by permutation and/or sign changes of the components in the initial vector. The components of the initial vector are assumed to be non-negative.

4.1.2 ML detection

One nice property with Permutation Modulation is that the maximum-likelihood detector is easy to describe and to instrument. We assume that the components of the initial vector are ordered so that:

$$\mu_1 < \mu_2 < ... < \mu_k$$

The maximum-likelihood detector for Variant I produces an estimate of the transmitted vector by replacing the m_1 smallest components with μ_1, the m_2 next smallest with μ_2, etc. The detector for Variant II works with the magnitudes of the components of the received vector, makes the estimation as in Variant I and then replaces the signs of the components in the estimate with the signs of the components of the received vector.

4.1.3 The initial vector problem

The initial vector which yields maximal d_{min} with given signal power S and given dimensionality n is known for Permutation Modulation. [39]

For both variants the difference $\mu_{i+1} - \mu_i$ should be 1.

For Variant I the number of components m_i should be chosen as:

$$m_i = \text{int } e^{-\frac{\eta + \mu_i^2}{\lambda}}$$

For Variant II the result is if $\mu_1 = 0$:

$$m_i = \text{int } e^{-\frac{\eta + \mu_i^2}{\lambda}} \qquad \text{for } i>1 \quad \text{and} \quad m_1 = \text{int } \frac{1}{2} e^{-\frac{\eta}{\lambda}}$$

and $\qquad m_i = int\ e^{-\dfrac{\eta + \mu_i^2}{\lambda}} \qquad$ if $\qquad \mu_1 > 0$

Here int x denotes the integer value of x.

Note that the amplitude distribution of the components in the initial vector is sampled Gaussian.

4.1.4 Performance

The performance of Permutation Modulation with the above choice of initial vector has been calculated. The error probability may be calculated using the close approximations in [4]. Some results are depicted in Figure 10. Here $\underline{m} = m_1, \dots, m_k$. The rate R is in bits/dim.

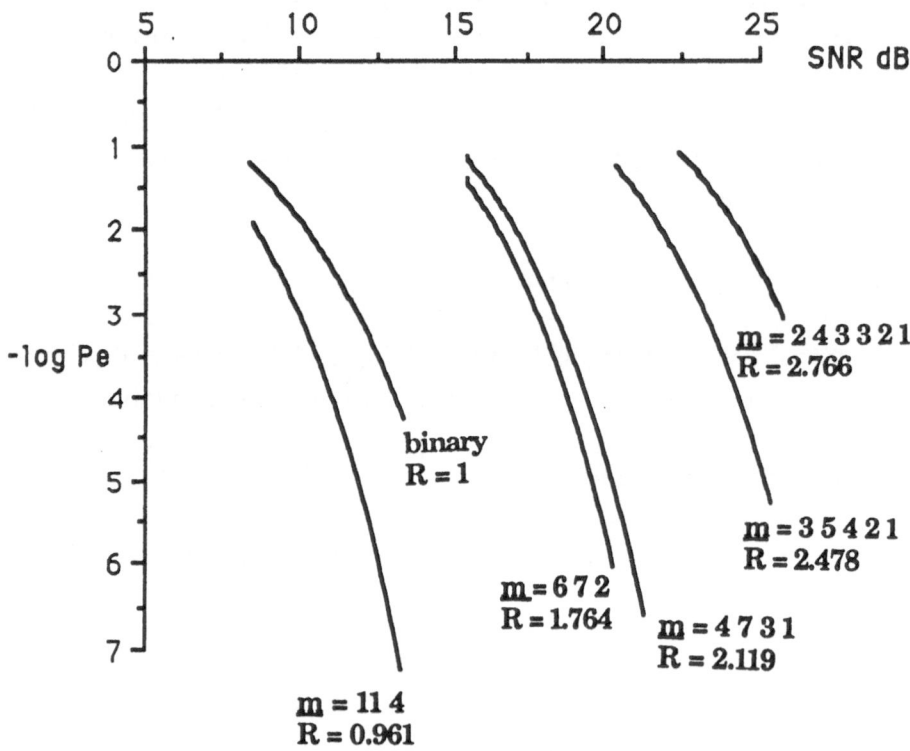

Figure 10 The error probability for permutation modulation and (for comparison) for antipodal binary signals

4.2 Other Non-commutative Group Codes

Besides Permutation Modulation there are just two other types of Non-Commutative Group Codes published in the open literature. One is the fourdimensional codes based on quarternions published by Zetterberg and Brändström [28] and the other is the combination of Commutative Group Codes with permutations published by Ottoson [13], [15]. Both are conveniently described by their projections onto orthogonal two-dimensional planes, as Commutative Group Codes were described in Figure 9. Since the Non-commutative Group Codes are not real-reducible to a direct sum of twodimensional subcodes they can not be projected onto single circles in the orthogonal planes. We are still able, though, to use a few circles. This suggests a simple representation of the codes as signals with phase and amplitude modulation in a few discrete levels.

Some of codes described in [28] are depicted in Figures 11 and 12.

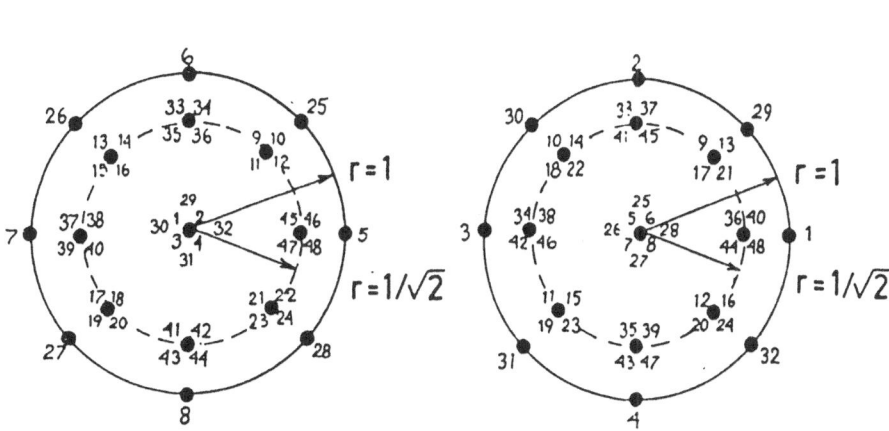

Figure 11 Group Codes generated by the quaternion group (points 1 - 8), the binary tetrahedral group (points 1 - 24) and the binary octahedral group (points 1 - 48)

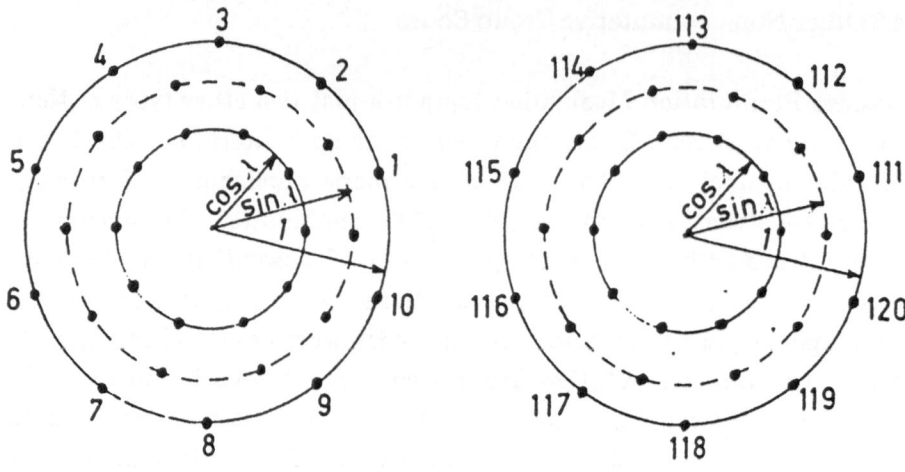

Figure 12 The Group Code generated by the binary icosahedral group

5. APPLICATION TO SPECIFIC CHANNELS

The Group Codes were initially intended to be used on a communication channel which could be modelled as a Gaussian channel. They have proved to be useful, however, also on other channels. In this section we will cover the published applications to other than the Gaussian channel.

In addition, as we have mentioned, Permutation Modulation has been used for source coding.

5.1 Fading Channels

Gaarder [16] has analyzed binary permutation modulation, Variant I, used on a fast fading Gaussian channel. The signals are orthogonal narrowband signals. The binary symbols are represented by a given signal for a binary one and no signal for a binary zero. Each orthogonal component is disturbed by independent multiplicative and additive Gaussian noise. The maximum-likelihood receiver is the same as for the non-fading channel: the signals with the largest envelopes are estimated as representatives of binary ones.

5.2 Discrete Channels

Blake [23] discussed codes defined as permutations of the components of a given vector over GF(q). The metric used is the Hamming metric. For sharply k-transitive groups of permutations the minimum Hamming distance is derived and found to be comparable to Reed-Solomon codes with approximately the same parameters. Very few k-transitive groups exist, however. The decoding problem is solved in some special cases.

To find other useful permutation groups in this context is still an open problem.

5.3 Broadcast Channels

The objective of the transmitter in a Gaussian broadcast channel is to transmit messages to two (or more) receivers disturbed by noise of different variance. This can be achieved by using a set of Group Codes generated by a subgroup of a larger group of orthogonal matrices and cosets to the subgroup. The basic reason for this is that the properties of a Group Code remains unchanged if every vector in the code is multiplied with the same orthogonal matrix. Thus the properties of the Group Code generated by the subgroup is the same as the properties of the code generated by the left cosets to the subgroup.

More formally let G be a (proper) subgroup of S, a group of orthogonal matrices. The elements of S may be partitioned into cosets $S_i G$ with i = 0, ... ,$(\| S \| / \| G \|) - 1$. S_0 is the unit matrix. The codes $S_i G X_1$ are orthogonal transformations of the Group Code $G X_1$. These codes are called clouds. The union of the clouds form a Group Code generated by S. This code is used to transmit messages to the receiver with least noise. The other receiver has higher noise level and can only decide which cloud the transmitted message belonged to.

This idea was first used by Heegard, dePedro and Wolf [30]. They used Permutation Modulation. The generalization to general group codes was made by Downey and Karlof [34] who later made a further generalization to more than two receivers [36]. To do that they used a set of nested groups.

5.4 Jamming Channels

Codes generated by a group with cosets was also used by Ericson [38] on the jamming channel. Such a channel is subject to active interference from an enemy which tries to prevent efficient transmission on the channel. The idea used by Ericson is similar to encryption. The vectors in a Group Code is partitioned into parts, similar to the clouds in the codes used on the broadcast channel as described above. One part is randomly selected. This corresponds to the choice of an encryption key. Only the legitimate receiver knows the key, i e the which partition is selected. The message to be sent is then represented by one of the code vectors in the selected partition.

Ericson uses a normal subgroup G to a larger group S of orthogonal matrices. Since the subgroup is normal (selfconjugate) the cosets form a group. A particular coset, say $S_i G$, is chosen by the key. The jammer, who does not know S_i, is faced with the problem of jamming any one of the code vectors in the Group Code generated by S. The legitimate receiver, who knows S_i, has to detect a vector in the smaller code generated by the coset $S_i G$.

6. REFERENCES

The references are listed in order of publication date.

[1]   D. Slepian: Large Signalling Alphabets Generated by Groups. Bell Telephone Laboratories, unpublished memo 1951.

[2]   C.E. Shannon: Probability of Error for Optimal Codes in a Gaus sian Channel. Bell System Tech. J., vol. 38, May 1959, pp 611 - 656.

[3]   D. Slepian: Bounds on Communication. Bell System Tech. J., vol. 42, May 1963, pp 681 - 707.

[4]   D. Slepian: Permutation Modulation. Proc. IEEE, vol. 53, March 1965, pp 228 - 236.

[5]   J.M. Wozencraft and I.M. Jacobs: Principles of Communication Engineering. New York: Wiley, 1965.

[6]   J.G. Dunn: Coding for Continuous Sources and Channels. Ph. D. Thesis, Dep. Elec. Eng., Columbia University, New York, May 1965.

[7]   W.W. Peterson and T. Kasami: Reliability Bounds for Polyphase Codes for the Gaussian Channel. Dep. of Elec. Eng., University of Hawaii, Scientific Report No. 3, July 1965.

[8]   L.H. Zetterberg: A Class of Codes for Polyphase Signals on a Bandlimited Gaussian Channel. IEEE Trans. on Information Theory, vol. IT-11. July 1965, pp 385 - 395.

[9]   L.H. Zetterberg: Detection of a Class of Coded and Phase-Modulated Signals. IEEE Trans. on Information Theory, vol. IT-12, April 1966, pp 153 - 161.

[10]   G. Einarsson: Performance of Polyphase Signals on a Gaussian Channel. Ericsson Technics, no.4, 1967, pp 411 - 433.

[11]   G.Einarsson: Polyphase Coding for a Gaussian Channel. Ericsson Technics, no.2, 1968, pp 75 - 130.

[12]   D.Slepian: Group Codes for the Gaussian Channel. Bell System Tech. J., vol. 47, April 1968, pp. 575 - 602.

[13]   R.Ottoson: Performance of Phase- and Amplitude- Modulated Signals on a Gaussian Channel. Ericsson Technics, no.3, 1969, pp 153 - 198.

[14]   I.Ingemarsson: Commutative Group Codes for the Gaussian
       Channel. Ph. D. Thesis, Dep. of Elec. Eng., Royal Inst. of
       Technology, Stockholm, Sweden,1970.

[15]   R.Ottoson: Group Codes for Phase- and Amplitude- Modulated
       Signals on a Gaussian Channel. IEEE Trans. on Information
       Theory, vol. IT-17, May 1971, pp 315 - 321.

[16]   N.T. Gaarder: Probability of Error for Binary Permutation
       Modulation on a Fading Gaussian Channel. IEEE Trans. on
       Information Theory, vol. IT-17, July 1971, pp 412 - 418.

[17]   D. Slepian: On Neighbor Distances and Symmetry in Group
       Codes. IEEE Trans. on Information Theory, vol. IT-17, Sept. 1971,
       pp 630 - 632.

[18]   D.Z. Djokovi´c and I.F. Blake: An Optimization Problem for Uni
       tary and Orthogonal Representations of Finite Groups. Trans.
       Amer. Math. Soc., vol. 164, Feb. 1972, pp 267 - 274.

[19]   E. Biglieri and M. Elia: On the Existence of Group Codes for the
       Gaussian Channel. IEEE Trans. on Information Theory, vol. IT-
       18, May 1972, pp 399 - 402.

[20]   I.F. Blake: Distance Properties of Group Codes for the Gaussian
       Channel. SIAM J. Appl. Math., vol.23, 1972, pp 312 - 324.

[21]   I. Ingemarsson: Commutative Group Codes for the Gaussian
       Channel. IEEE Trans. on Information Theory, vol. IT-19, March
       1973, pp 215 - 219.

[22]   I.F. Blake: Configuration Matrices of Group Codes. IEEE Trans.
       on Information Theory, vol. IT-20, Jan. 1974, pp 95 - 100.

[23]   I.F. Blake: Permutation Codes for Discrete Channels. IEEE
       Trans. on Information Theory, vol. IT-20, Jan. 1974, pp 138 - 140.

[24]   I.F. Blake and R.C. Mullin: The Mathematical Theory of Coding.
       New York: Academic Press, 1975, chapter 6.

[25]   E.M. Biglieri and M. Elia: Cyclic-Group Codes for the Gaussian
       Channel. IEEE Trans. on Information Theory, vol. IT-22, Sept.
       1976, pp 624 -629.

[26]   E.M. Biglieri and M. Elia: Optimum Permutation Modulation
       Codes and Their Asymptotic Performance. IEEE Trans. on
       Information Theory, vol. IT-22, Nov. 1976, pp 751 - 753.

[27]   C.P. Downey and J.K. Karlof: On the Existence of [M,n] Group
       Codes for the Gaussian Channel with M and n Odd. IEEE Trans.
       on Information Theory, vol. IT-23, July 1977, pp 500 - 503.

[28]   L.H. Zetterberg and H. Brändström: Codes for Combined Phase
        and Amplitude Modulated Signals in a Four-Dimensional Space.
        IEEE Trans. on Communication, vol. COM-25, Sept. 1977, pp 943 -
        950.

[29]   J.K. Karlof and C.P. Downey: Odd Group Codes for the Gaussian
        Channel. SIAM J. Appl. Math., vol. 34, June 1978, pp 715 - 716.

[30]   C. Heegard, H.E. DePedro and J.K. Wolf: Permutation Codes for
        the Gaussian Broadcast Channel with two Receivers. IEEE
        Trans. on Information Theory, vol. IT-24, Sept. 1978, pp 569 - 578.

[31]   C.P. Downey and J.K. Karlof: Optimal [M,3] Group Codes for the
        Gaussian Channel. IEEE Trans. on Information Theory, vol. IT-
        24, Nov. 1978, pp 760 - 761.

[32]   D.E. Lasi´c: Class of Block Codes for the Gaussian Channel.
        Electronic Letters, vol. 16, Feb. 1980. pp 185 - 186.

[33]   C.P. Downey and J.K. Karlof: Computation Methods for Optimal
        [M,3] Group Codes for the Gaussian Channel. Utilitas
        Mathematica, vol. 18, 1980, pp 51 - 70.

[34]   C.P. Downey and J.K. Karlof: Group Codes for the Gaussian
        Broadcast Channel with two Receivers. IEEE Trans. on
        Information Theory, vol. IT-26, July 1980, pp 406 - 411.

[35]   D.E. Lasi´c, D.B. Draji´c and V. Senk: A Table of Some Small-Size
        Three-Dimensional Best Spherical Codes. Presented at the IEEE
        Int. Symp. on Information Theory, Les Arcs, France, 1982.

[36]   C.P. Downey and J.K. Karlof: Group Codes for the M-Receiver
        Gaussian   Broadcast Channel. IEEE Trans. on Information
        Theory, vol. IT-29, July 1983, pp 595 - 597.

[37]   V.V. Ginzburg: Multidimensional Signals for a Continuous
        Channel. Problemy Peredachi Informatsii, vol. 20, Jan - March
        1984, pp 28 - 46.

[38]   T. Ericson: A Min-Max Theorem for Antijamming Group Codes.
        IEEE Trans. on Information Theory, vol It- 30, Nov. 1984, pp 792 -
        799.

[39]   I. Ingemarsson: Optimized Permutation Modulation. Submitted
        to IEEE Trans. on Information Theory.

Distances and distance bounds for convolutional codes—an overview

Rolf Johannesson*   Kamil Sh. Zigangirov**

* Department of Information Theory
University of Lund
Box 118
S-221 00 Lund, Sweden

** Institute for Problems of Information Transmission
USSR Academy of Sciences
19 Ermolovoy st.
Moscow GSP-4, USSR 101447

Abstract—In this paper we give a self-contained overview of known distance measures for convolutional codes and of upper and lower bounds on the free distance. The upper bounds are valid for general trellis codes and for convolutional codes, respectively. The lower bound is valid for time-varying convolutional codes. We also present a new lower bound on the distance profile for fixed convolutional codes.

1   Introduction

Over the past decade there has been a significant increase in using both maximum-likelihood (Viterbi) decoding and sequential decoding to achieve reliable communication. We can expect that the demand for communication with low or extremely low error probability will continue to grow. It is well-known [Vit71] [MaC71] that the free distance is the principal determiner of the event error probability when maximum-likelihood (or nearly so) decoding is used for a convolutional code. It has also been observed [MaC71] [Joh75] [ChC76] that an optimum distance profile is desirable to obtain a good computational performance with sequential decoding. Thus, it is important to find methods for constructing convolutional codes with both a large free distance and a good distance profile. The existence of good upper and lower bounds makes such methods more effective.

In Section 2 we define both general trellis encoders and convolutional encoders. The ensembles of time-varying and fixed convolutional codes are discussed. Several distance measures are defined and discussed in Section 3. In

Section 4 we present upper bounds on the free distance for general trellis codes and fixed convolutional codes and in Section 5 we give a new proof of Costello's lower bound on the free distance of time-varying convolutional codes. A new lower bound on the distance profile for fixed convolutional codes is presented in Section 6.

2 Structural properties

In a rate $R = b/c$, binary, general trellis encoder the information sequence

$$i_0 i_1 i_2 \ldots = i_0^{(1)} i_0^{(2)} \ldots i_0^{(b)} i_1^{(1)} i_1^{(2)} \ldots i_1^{(b)} i_2^{(1)} i_2^{(2)} \ldots i_2^{(b)} \ldots$$

is encoded as the sequence

$$t_0 t_1 t_2 \ldots = t_0^{(1)} t_0^{(2)} \ldots t_0^{(c)} t_1^{(1)} t_1^{(2)} \ldots t_1^{(c)} t_2^{(1)} t_2^{(2)} \ldots t_2^{(c)} \ldots ,$$

where

$$t_j = f_j(i_j, i_{j-1}, \ldots, i_{j-m}). \tag{1}$$

The parameter m is called the code *memory*. The function f_j is in general time-dependent. In Fig. 1 we show a general trellis encoder.

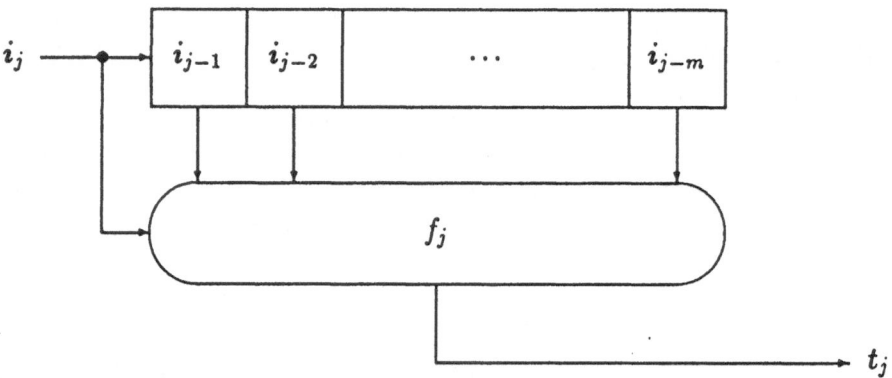

Figure 1: A general trellis encoder.

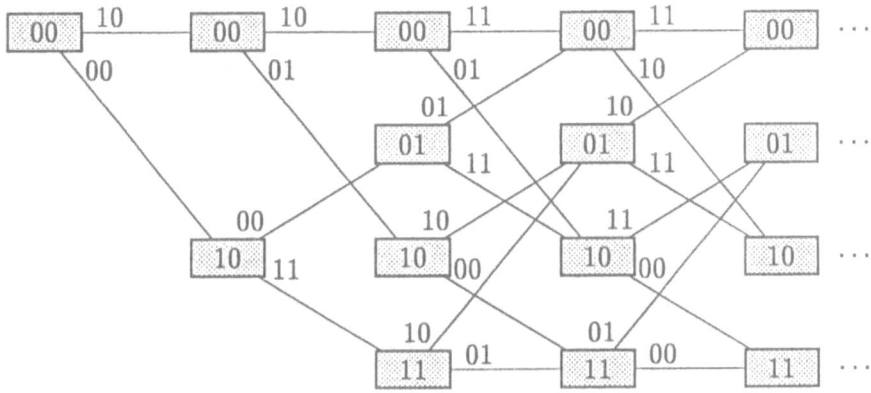

Figure 2: A rate $R = 1/2$ trellis code.

The state diagram for our encoder resembles a trellis. (The term *trellis* was coined by Forney [For67].) In Fig. 2 we show a trellis diagram for a rate $R = 1/2$, memory $m = 2$, binary trellis code.

Starting at the leftmost 00-state at time 0 we traverse the trellis from left to right and leave a state using the upper branch when the corresponding information symbol is 0 and the lower branch when it is 1. Thus the information sequence 1001... is encoded as the code sequence 00000110....

A rate $R = b/c$, binary *convolutional* encoder is a *linear*, binary trellis encoder, i.e. the functions $f_j, j = 0, 1, 2, \ldots$, are required to be linear functions from $GF(2)^{(m+1)b}$ to $GF(2)^c$. It is often convenient to represent such functions as

$$t_j = i_j G_0(j) + i_{j-1} G_1(j) + \ldots + i_{j-m} G_m(j), \qquad (2)$$

where each $G_i(j), i = 0, 1, \ldots, m$, is a binary $b \times c$ matrix and the additions are calculated positionwise modulo 2. The matrices are in general time-dependent. In Fig. 3 we show a general time-varying convolutional encoder.

We call an ensemble of trellis codes *path-independent* if the digits on any path diverging from the all zero path are mutually independent over the span to the next node (if any) where the path remerge with the all zero path.

Theorem 1: The ensemble of binary, time-varying convolutional codes of rate $R = b/c$ and memory m in which each digit in each of the matrices $G_i(j)$ for $0 \leq i \leq m$ and $j = 0, 1, 2, \ldots$ is chosen independently with probability 1/2 is path-independent and all code symbols on a span differing from the all zero path are equally likely to be 0 and 1. □

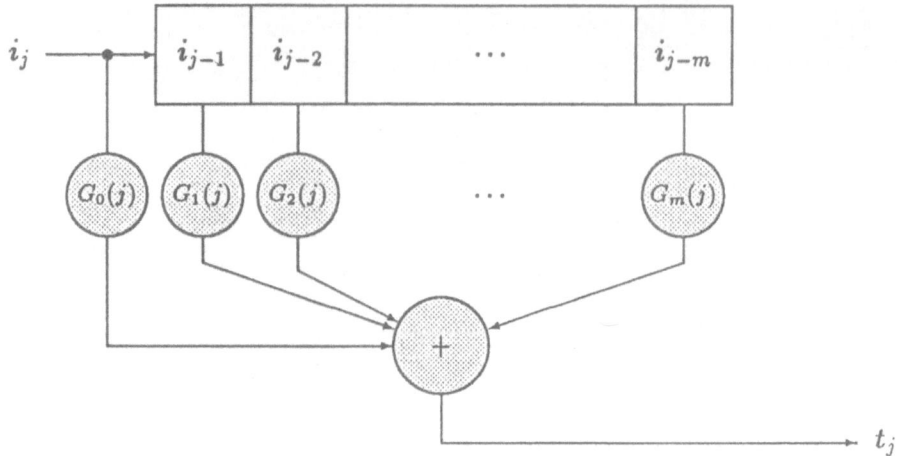

Figure 3: A general time-varying convolutional encoder.

Proof: Suppose that a path diverges from the all zero path at depth u and that these paths remain unmerged until at least depth v. If $i_j, u \leq j \leq v$, is the corresponding information sequence, the encoded sequence over the diverging path is given by (2), and at least one of the information tuples $i_{j-i}, 0 \leq i \leq m$, say $i = k$, is $\neq \mathbf{0}$. Thus, regardless of the choice of the matrices $G_i(j), i \neq k$, the tuple $t_j, u \leq j \leq v$, over all choices of $G_k(j)$ will take on all values with the same probability. $\qquad \square$

In Section 5 we shall derive a lower bound on the free distance for this ensemble of time-varying convolutional codes.

When the matrices $G_i(j)$ in (2) are all time-invariant we have a binary, *fixed* or *time-invariant* convolutional encoder (FCE) and (2) becomes

$$t_j = i_j G_0 + i_{j-1} G_1 + \ldots + i_{j-m} G_m, \tag{3}$$

where $G_i, 0 \leq i \leq m$, is a binary $b \times c$ matrix.

In Fig. 4 we show a general fixed convolutional encoder. For fixed convolutional encoders it is sometimes—as an alternative to the trellis diagram—useful to draw the state diagram as a de Bruijn graph [Gol67].

In Fig. 5 we show such a graph for the rate $R = 1/2$, memory $m = 2$, binary, fixed convolutional encoder with $G_0 = (11), G_1 = (10)$, and $G_2 = (11)$.

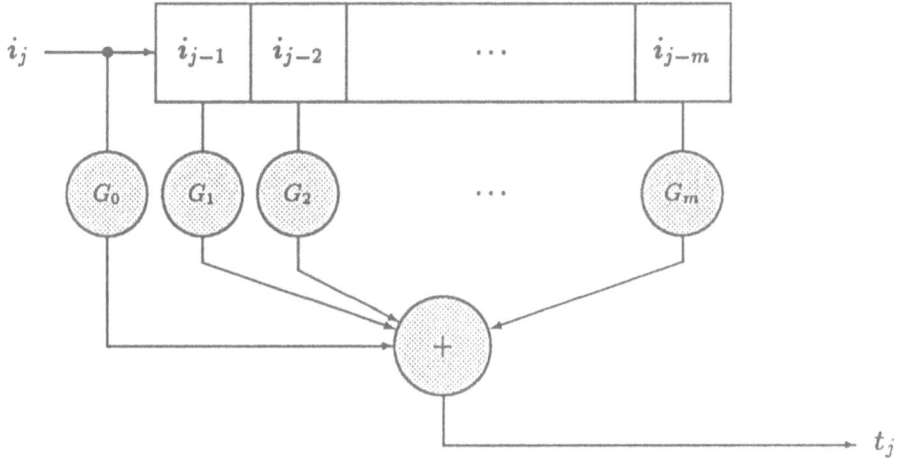

Figure 4: A general fixed convolutional encoder.

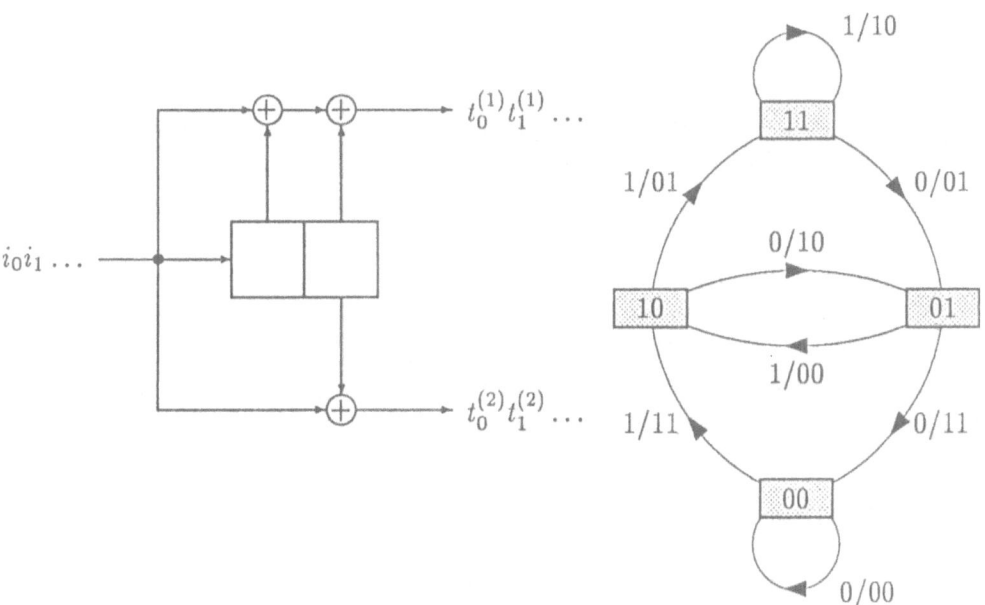

Figure 5: A rate $R = 1/2$ fixed convolutional encoder and its state diagram.

Using (3) we can rewrite the expression for the code sequence as

$$t_0 t_1 t_2 \ldots = (i_0 i_1 i_2 \ldots)G, \tag{4}$$

where

$$G = \begin{pmatrix} G_0 & G_1 & G_2 & \ldots & G_m & & & & \\ & G_0 & G_1 & G_2 & \ldots & G_m & & & \\ & & G_0 & G_1 & G_2 & \ldots & G_m & & \\ & & & \ddots & \ddots & \ddots & & \ddots & \\ & & & & G_0 & G_1 & G_2 & \ldots & G_m \\ & & & & & \ddots & \ddots & \ddots & & \ddots \end{pmatrix}. \tag{5}$$

Theorem 2: In the ensemble of binary, fixed convolutional codes of rate $R = b/c$ and memory m in which each digit in each of the generator matrices G_i for $0 \leq i \leq m$ is chosen independently with probability $1/2$ the first $(m+1)c$ code symbols on a path diverging from the all zero path are independent and equally likely to be 0 and 1. □

Proof: Assume that a path diverge from the all zero path at depth j_0, i.e. $i_{j_0} \neq 0$. Then we have

$$t_{j_0+j} = i_{j_0+j}G_0 + i_{j_0+j-1}G_1 + \ldots + i_{j_0}G_j + \ldots + i_{j_0+j-m}G_m, \tag{6}$$

where $0 \leq j \leq m$. Since t_{j_0+j} is equal to a fix vector determined by $G_0, G_1, \ldots,$ G_{j-1} plus $i_{j_0}G_j$, where $i_{j_0} \neq 0$ and G_j is equally likely to be any $b \times c$ binary matrix, t_{j_0+j} assumes each of its 2^c possible values with the same probability. Furthermore, it is independent of the previous code symbols $t_{j_0}, t_{j_0+1}, \ldots, t_{j_0+j-1}$ for $1 \leq j \leq m$ and the proof is complete.

□

In Section 6 we shall prove a lower bound for the ensemble of binary, fixed convolutional codes in which each digit in each of the generator matrices is chosen independently to be zero and one with probability $1/2$.

3 Distance measures

Several distance measures have been proposed for trellis codes. In this section we discuss the most important ones.

We shall find it convenient to write

$$t_{[0,n]} = t_0 t_1 \ldots t_n = t_0^{(1)} t_0^{(2)} \ldots t_0^{(c)} t_1^{(1)} t_1^{(2)} \ldots t_1^{(c)} \ldots t_n^{(1)} t_n^{(2)} \ldots t_n^{(c)}$$

for the encoded path containing the first $n+1$ "branches" of the encoded sequence.

The encoded path $t_{[0,m]}$ is called the *first constraint path* and its length, $n_c = (m+1)c$, is called the *constraint length* of the code.

The most fundamental distance measure is called the *column distance* [Cos69] [Mas75]:

Definition: The jth order *column distance* d_j for a time-invariant trellis code is the minimum Hamming distance between two encoded sequences $t_{[0,j]}$ resulting from information sequences $i_{[0,j]}$ with differing i_0. $\quad\square$

For *linear*, time-invariant trellis codes, i.e. fixed convolutional codes, we have that d_j is also the minimum of the Hamming weights of the paths $t_{[0,j]}$ resulting from information sequences with $i_0 \neq 0$. Thus,

$$d_j = \min_{i_0 \neq 0} w_H(t_0 t_1 \ldots t_j), \tag{7}$$

where $w_H(\ )$ denotes the Hamming weight of a sequence. Let G_j^c denote the truncation of the semi-infinite super-matrix G after $j+1$ *supercolumns*, i.e.

$$G_j^c = \begin{pmatrix} G_0 & G_1 & G_2 & \cdots & G_j \\ & G_0 & G_1 & & G_{j-1} \\ & & G_0 & & G_{j-2} \\ & & & \ddots & \vdots \\ & & & & G_0 \end{pmatrix}, \tag{8}$$

where $G_i = 0$ when $i > m$. Making use of (3) we can rewrite (7) as

$$d_j = \min_{i_0 \neq 0} w_H((i_0 i_1 \ldots i_j) G_j^c). \tag{9}$$

From (8) it follows that to obtain the jth column distance d_j we truncate the super-matrix G after $j+1$ supercolumns.

The quantity d_m is called the *minimum distance* of the code and determines the error correcting capability of a decoder that estimates the information symbol i_0 based the received symbols over the first constraint length only.

A good computational performance for sequential decoding requires a rapid initial growth of the column distances [MaC71]. This led to the introduction of the $(m+1)$-tuple

$$d = [d_0, d_1, \ldots, d_m], \tag{10}$$

which is called the *distance profile* [Joh75].

A code is said to have a distance profile d *superior* to a distance profile d' of another code of the same rate R and memory m, when there is some l such that

$$d_j \begin{cases} = d'_j, & j = 0, 1, \ldots, l-1 \\ > d'_j, & j = l. \end{cases} \tag{11}$$

Moreover, a code is said to have an *optimum distance profile* (ODP) if there exists no code of the same rate and memory with a better distance profile. An ODP code has the fastest possible initial growth of the minimal separation between paths diverging at the root.

The column distance d_j is a non-decreasing function of j and is sometimes called the *column distance function* [ChC76]. Since it is also finite the limit

$$d_\infty = \lim_{j \to \infty} d_j \qquad (12)$$

exits and we have the relations

$$d_0 \leq d_1 \leq d_2 \leq \ldots \leq d_\infty. \qquad (13)$$

The quantity d_∞ is the minimum Hamming distance between any two paths of infinite length with a differing first information symbol. This led Massey to introduce the terminology *free distance* for this distance measure which is the principal determiner for the error correcting capability of a code when maximum likelihood (or nearly so) decoding is used.

Often we need a more detailed knowledge of the distance structure of the code. For a fixed convolutional code we let $n(d_\infty + i)$ denote the number of weight $d_\infty + i$ paths which depart from the all zero path at the root in the code trellis and do not reach the zero state until their termini. We call $n(d_\infty + i)$ the $(i+1)$-th *spectral component*. The sequence

$$n(d_\infty + i), i = 0, 1, 2, \ldots, \qquad (14)$$

is called the *distance spectrum* of the code.

As a counterpart to the column distance we have the *row distance* [Cos69] [Mas75]: .

Definition: The jth order *row distance* r_j is the minimum Hamming distance between two paths $t_{[0,j+m]}$ and $t'_{[0,j+m]}$ which diverge at some node (not necessarily the root) and remerge at some node. □

By the linearity of fixed convolutional codes r_j is also the minimum of the Hamming weights of the paths $t_{[0,j+m]}$ resulting from information sequences $i_{[0,j]} \neq 0$. Thus we have

$$r_j = \min_{i_{[0,j]} \neq 0} w_H((i_0 i_1 \ldots i_j) G_j^r), \qquad (15)$$

where the matrix G_j^r is formed by truncating the semi-infinite super-matrix G after its first $j+1$ *superrows*, i.e.

$$G_j^r = \begin{pmatrix} G_0 & G_1 & \ldots & G_m & & & \\ & G_0 & G_1 & \ldots & G_m & & \\ & & G_0 & G_1 & \ldots & G_m & \\ & & & \ddots & & & \ddots \\ & & & & G_0 & G_1 & \ldots & G_m \end{pmatrix}. \qquad (16)$$

From (16) it follows that the row distance r_j quite contrary to the column distance d_j is a non-increasing function of j. Since every r_j is also non-negative the limit

$$r_\infty = \lim_{j \to \infty} r_j \tag{17}$$

exits and we have the relations

$$r_\infty \leq \ldots \leq r_2 \leq r_1 \leq r_0. \tag{18}$$

If we think of the row distances in terms of a state diagram it follows that r_0 is the weight of the shortest path diverging from and returning to the zero state. Higher order row distances are obtained by allowing successively more freedom in finding the minimum weight path diverging from and returning to the zero state. The jth order row distance r_j is the minimum weight of any path of length $j + m + 1$ branches diverging from and returning to the zero state. Eventually, r_∞ is the minimum weight path diverging from and returning to the zero state. Since the column distance d_i is the minimum weight of a path of length $i + 1$ branches and with a first branch diverging from the zero state it is obvious that

$$d_i \leq r_j, \text{ all } i \text{ and } j, \tag{19}$$

and thus that

$$d_0 \leq d_1 \leq \ldots \leq d_i \leq \ldots \leq d_\infty \leq r_\infty \leq \ldots \leq r_j \leq \ldots \leq r_1 \leq r_0. \tag{20}$$

Furthermore, if there are no closed loops of zero weight in the state diagram except the trivial zero weight self-loop at the zero state, it follows that

$$d_\infty = r_\infty. \tag{21}$$

The existence of a zero weight nontrivial loop in the state diagram is equivalent to the existence of an information sequence with infinite weight that is encoded as a finite weight sequence. Thus, a finite number of errors in the code sequence can convert it to the all zero sequence which, of course, will be decoded as the all zero information sequence resulting in an infinite number of errors in the decoded information symbols—a catastrophic situation! Such an encoder is called *catastrophic* and should be avoided.

It can be shown that a fixed convolutional encoder is catastrophic if and only if it is not feedforward-invertible, i.e. it has no inverse which is a feedforward linear sequential circuit [MaS68][For70]. For a non-catastrophic encoder equality (21) holds.

Since all fixed convolutional codes used in practice are non-catastrophic the row distance could probably blame equality (21) for not getting much attention in the literature. However, its significance should not be underestimated. It is easy to calculate and serves as an excellent rejection rule when codes are tested in a search for codes with large free distance.

Example 1:

The state diagram for the following rate $R = 1/2$, binary, fixed convolutional encoder of memory $m = 3$

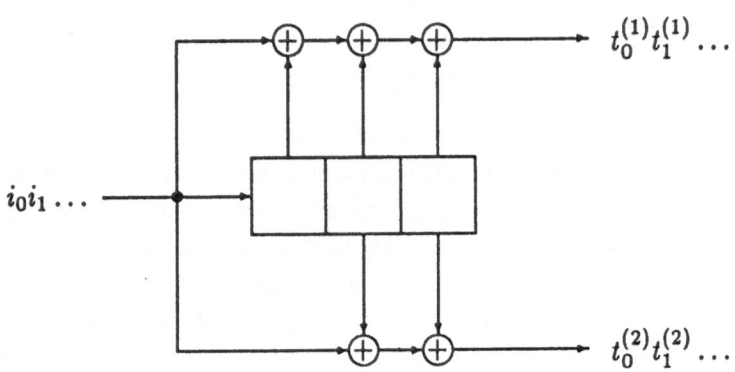

is shown below:

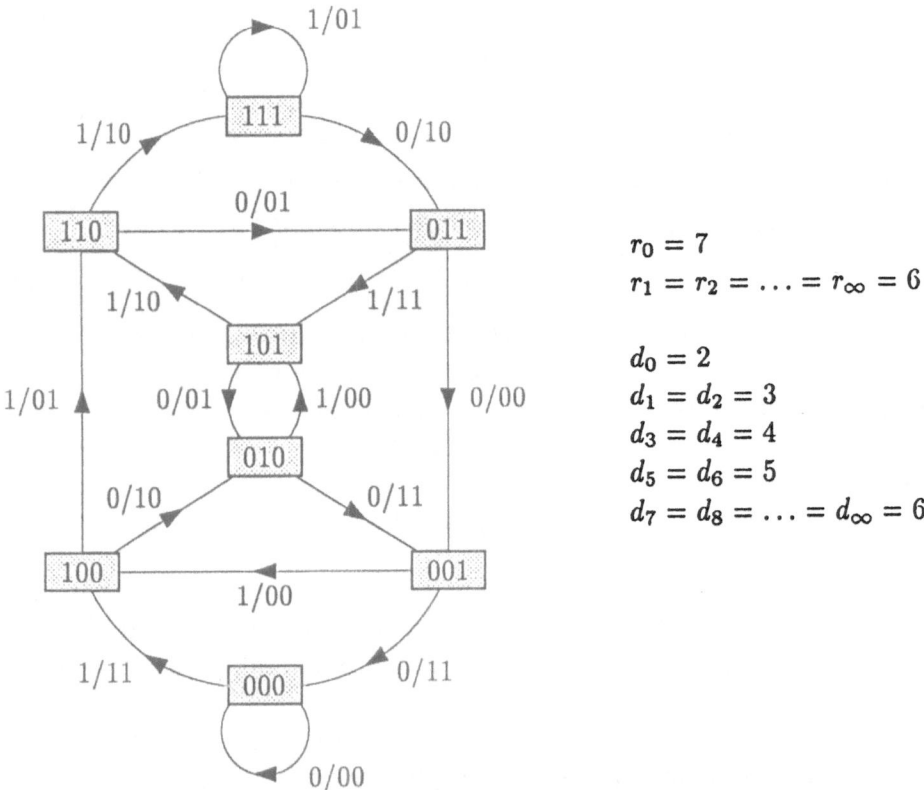

$r_0 = 7$
$r_1 = r_2 = \ldots = r_\infty = 6$

$d_0 = 2$
$d_1 = d_2 = 3$
$d_3 = d_4 = 4$
$d_5 = d_6 = 5$
$d_7 = d_8 = \ldots = d_\infty = 6$

The row distance r_0 is obtained from the loop $000 \to 100 \to 010 \to 001 \to 000$, and $r_1, r_2, \ldots, r_\infty$ are obtained from the loop $000 \to 100 \to 110 \to 011 \to$

$001 \rightarrow 000$. The column distance d_0 is obtained from $000 \rightarrow 100, d_1$ from e.g. $000 \rightarrow 100 \rightarrow 110, d_2$ from $000 \rightarrow 100 \rightarrow 010 \rightarrow 101$, etc.

The row and column distances are shown below:

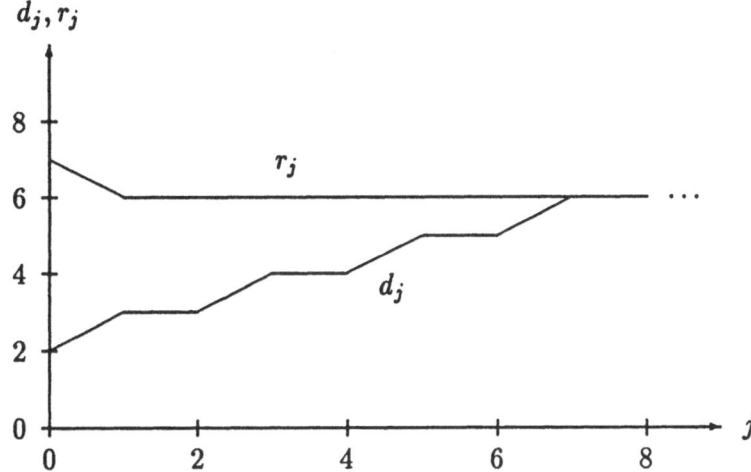

We have the free distance $d_\infty = r_\infty = 6$ and the distance profile $d = [2, 3, 3, 4]$. This encoder is ODP and has optimum free distance (OFD). □

Example 2:

The following rate $R = 1/2$ fixed convolutional encoder of memory $m = 3$

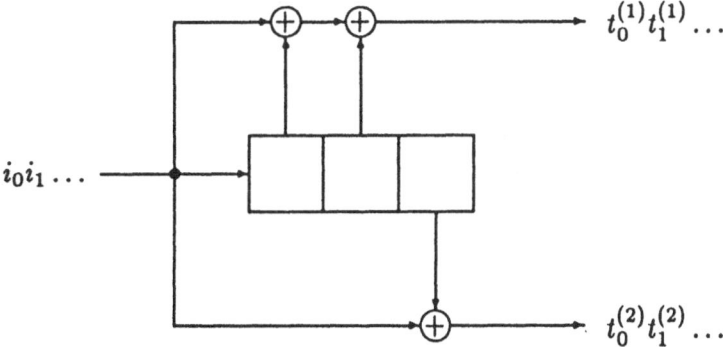

is catastophic, since its state diagram has a non-trivial closed zero weight loop, viz. $110 \rightarrow 011 \rightarrow 101 \rightarrow 110$:

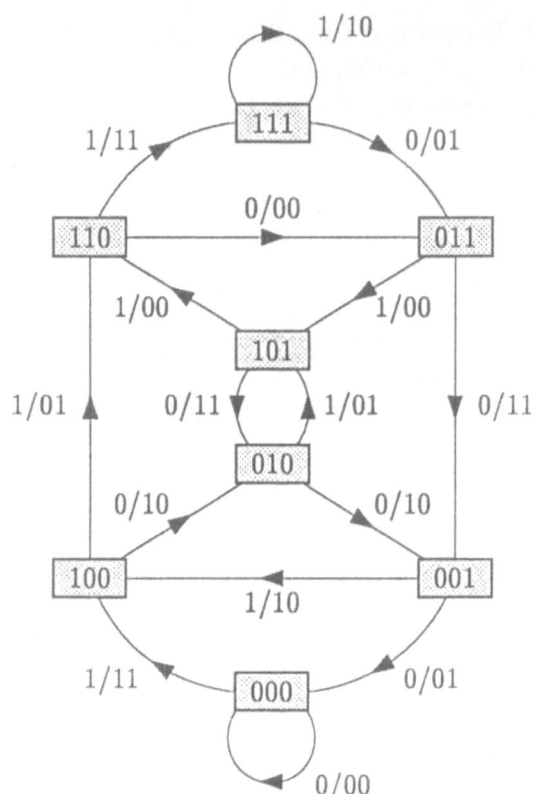

$$r_0 = r_1 = \ldots = r_\infty = 5$$

$$d_0 = 2$$
$$d_1 = d_2 = \ldots = d_\infty = 3$$

The row distances $r_0 = r_1 = \ldots = r_\infty$ are obtained from the loop $000 \rightarrow 100 \rightarrow 010 \rightarrow 001 \rightarrow 000$. The column distance d_0 is obtained from $000 \rightarrow 100, d_1$ from $000 \rightarrow 100 \rightarrow 110, d_2$ from $000 \rightarrow 100 \rightarrow 110 \rightarrow 011$, etc. The distance d_∞ is obtained from $000 \rightarrow 100 \rightarrow 110 \rightarrow 011 \rightarrow 101 \rightarrow 110 \rightarrow \ldots \rightarrow 110 \rightarrow 011 \rightarrow 101 \rightarrow 110 \rightarrow \ldots$. The row distance $r_\infty = 5$ is strictly larger than the free distance $d_\infty = 3$.

The column and row distances are shown below:

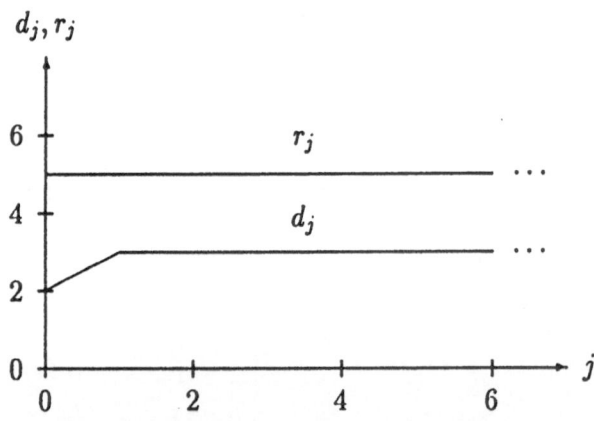

4 Upper bounds on free distance

We will now prove an upper bound on the free distance based on Plotkin's bound for block codes. It is valid for general trellis codes. For linear trellis codes we give a slightly better bound.

For the sake of completeness we start by proving Plotkin's upper bound on the minimum distance for block codes.

Lemma 3 (Plotkin bound for block codes): The minimum distance for any binary block code of M codewords and block length N satisfies

$$d_{\min} \leq \left\lfloor \frac{NM}{2(M-1)} \right\rfloor. \tag{22}$$

□

Proof: Consider an arbitrary column in the list of M code words. Suppose that the symbol 0 occurs n_0 times in this column. Its contribution to the sum of the distances between all ordered pairs of code words is $n_0(M - n_0)$. Since we have the same contribution from symbol 1, the total contribution from the column is

$$2\,n_0(M - n_0) \leq \frac{M^2}{2}, \tag{23}$$

with equality iff $n_0 = M/2$. Summing the distances over all N columns we have at most $NM^2/2$. Since $d_{\min}$ is the minimum distance between a pair of code words and since there are $M(M-1)$ ordered pairs, we have

$$M(M-1)d_{\min} \leq \frac{NM^2}{2}, \tag{24}$$

and the proof is complete. □

Heller [Hel68][LaM70] used Plotkin's bound for block codes to obtain a surprisingly tight bound on the free distance for trellis codes.

Theorem 4 (Heller bound for trellis codes): The free distance for any binary, rate $R = b/c$ trellis code of memory m satisfies

$$d_\infty \leq \min_{i \geq 1} \left\lfloor \frac{(m+i)c}{2(1 - 2^{-bi})} \right\rfloor. \tag{25}$$

□

Proof: In a memory m, binary, rate $R = b/c$ trellis there are totally 2^{bi} paths of length $(m+i)c$ symbols from the root to any node at depth $m+i, i = 1, 2, \ldots$.

These paths constitute a block code with $M = 2^{bi}$ code words and block length $N = (m + i)c$. Apply Lemma 3 for $i = 1, 2, \ldots$. □

Heller's bound is valid for any nonlinear, time-varying trellis code. For *linear*, time-constant convolutional codes we can use Griesmer's bound for block codes to obtain slight improvements for some memories [Gri60].

Lemma 5 (Griesmer bound for linear block codes): For a binary, linear, rate $R = K/N$ block code with minimum distance d_{min} we have

$$\sum_{i=0}^{K-1} \left\lceil \frac{d_{min}}{2^i} \right\rceil \leq N. \tag{26}$$

□

Proof: Without loss of generality we assume that the first row of the generator matrix G is $(111 \ldots 10 \ldots 0)$ with d_{min} ones. Every other row has at least $\lceil d_{min}/2 \rceil$ ones or $\lceil d_{min}/2 \rceil$ zeros in the first d_{min} positions. Hence they have at least $\lceil d_{min}/2 \rceil$ ones in the remaining $N - d_{min}$ positions. Therefore the residual code with respect to the first row is a rate $R = (K - 1)/(N - d_{min})$ code with minimum distance $\geq \lceil d_{min}/2 \rceil$. Using induction completes the proof. □

Consider all paths that diverge at the root and remerge at depth $m + i, i = 1, 2, \ldots$, in a binary, linear, rate $R = b/c$ trellis. The minimum distance for these $(N, K) = ((m + i)c, bi), i = 1, 2, \ldots$, linear block codes must satisfy the Griesmer bound. Hence we have

Theorem 6 (Griesmer bound for convolutional codes): The free distance for any binary, rate $R = b/c$ convolutional code of memory m satisfies

$$\sum_{j=0}^{bi-1} \left\lceil \frac{d_\infty}{2^j} \right\rceil \leq (m + i)c \tag{27}$$

for $i = 1, 2, \ldots$. □

Example 3:

a) Let $R = 1/2$ and $m = 16$. Since $\min_{i \geq 1} \lfloor (16 + i)/(1 - 2^{-i}) \rfloor = 21$ any binary, rate $R = 1/2$ trellis code with memory $m = 16$ must have $d_\infty \leq 21$.

Since $\sum_{j=0}^{3} \lceil 21/2^j \rceil = 41 \nleq (16 + 4)2 = 40$ any binary, rate $R = 1/2$ convolutional code with memory $m = 16$ must have $d_\infty < 21$. The Griesmer bound gives an improvement by one. In fact there exists such a code with $d_\infty = 20$ [CeJ88].

b) Let $R = 1/2$ and $m = 18$. Since $\min_{i \geq 1} \lfloor (18 + i)/(1 - 2^{-i}) \rfloor = 23$ any binary, rate $R = 1/2$ trellis code with memory $m = 18$ must have $d_\infty \leq 23$.

Since $\sum_{j=0}^{i-1}\lceil 23/2^j\rceil \leq (18+i)2$ for all $i \geq 1$ the Griesmer bound for convolutional codes does not give any improvement over the Heller bound for general trellis codes in this case. The largest free distance for any binary, rate $R = 1/2$ convolutional code with memory $m = 18$ has been determined by exhaustive search and is $d_\infty = 22$ [CeJ88]. The Griesmer bound is not tight. □

5 Lower bound on free distance

In this section we shall derive a lower bound on the free distance for the ensemble of random time-varying convolutional codes. This bound is due to Costello [Cos74] but our proof is slightly different. Our goal is to find a non-trivial upper bound on the probability that $d_\infty < d$, i.e. to prove that $P[d_\infty < d] < 1$, since then we know that there exists at least one code within our ensemble that has $d_\infty \geq d$.

Consider a time-varying convolutional code of rate $R = b/c$ and memory m in which each digit in each of the matrices $G_i(j)$ for $0 \leq i \leq m$ and $j = 0, 1, 2, \ldots$ is chosen independently with probability $1/2$. In the trellis there are $2^b - 1$ incorrect paths of length $m + 1$ branches, and at most $2^{(i-2)b}(2^b - 1)^2$ incorrect paths diverging at the root and first remerging at depth $m + i, i \geq 2$. We use 2^{ib} as an upper bound on the number of these incorrect paths of length $m + i, i \geq 1$. From Theorem 1 it follows that the probability for each of these paths is $2^{-(m+i)c}$, and that there are $\binom{(m+i)c}{j}$ ways of choosing exactly j ones among the $(m + i)c$ code symbols.

Using the union bound we can upper bound $P[d_\infty < d]$ by summing the probabilities for all incorrect paths with weights less than d. Thus, we have

$$P[d_\infty < d] < \sum_{j=0}^{d-1}\sum_{i=1}^{\infty} 2^{ib}\binom{(m+i)c}{j}2^{-(m+i)c}. \tag{28}$$

We use the substitution

$$k = (m+i)c \tag{29}$$

and upper bound (28) by summing over $k = 0, 1, 2, \ldots,$

$$P[d_\infty < d] < 2^{-mb}\sum_{j=0}^{d-1}\sum_{k=0}^{\infty}\binom{k}{j}2^{k(R-1)}. \tag{30}$$

Let

$$x = 2^{R-1}, \frac{1}{2} < x < 1, \tag{31}$$

and rearrange (30) as

$$P[d_\infty < d] \; < \; 2^{-mb} \sum_{j=0}^{d-1} \frac{x^j}{j!} \sum_{k=0}^{\infty} k(k-1)\ldots(k-j+1)x^{k-j}$$

$$= \; 2^{-mb} \sum_{j=0}^{d-1} \frac{x^j}{j!} \left(\sum_{k=0}^{\infty} x^k \right)^{(j)}$$

$$= \; 2^{-mb} \sum_{j=0}^{d-1} \frac{x^j}{j!} \left(\frac{1}{1-x} \right)^{(j)}$$

$$= \; 2^{-mb} \frac{1}{1-x} \sum_{j=0}^{d-1} \left(\frac{x}{1-x} \right)^{j}$$

$$= \; 2^{-mb} \frac{1}{1-\dot{x}} \frac{(\frac{x}{1-x})^d - 1}{\frac{x}{1-x} - 1}$$

$$< \; \frac{2^{-mb}}{(2x-1)(x^{-1}-1)^d}. \tag{32}$$

Using (31) we obtain

$$P[d_\infty < d] < \frac{2^{-mb}}{(2^R - 1)(2^{1-R} - 1)^d}. \tag{33}$$

Thus, if

$$\frac{2^{-mb}}{(2^{1-R} - 1)^d} \leq 2^R - 1, \tag{34}$$

then there exists at least one code of rate R and memory m with $d_\infty \geq d$.

Let $f(R)$ be a positive function of R such that

$$2^R - 1 = (2^{1-R} - 1)^{f(R)}. \tag{35}$$

Then by taking the logarithm of (34) we obtain

$$d \leq \frac{mb}{-\log_2(2^{1-R} - 1)} - f(R). \tag{36}$$

Finally, we take d equal to the right side of (36), and we have proved

Theorem 7 (Lower bound for time-varying convolutional codes): There exists a binary, time-varying convolutional code of rate $R = b/c$ and memory m with a free distance satisfying the inequality

$$\frac{d_\infty}{m} \geq \frac{b}{-\log_2(2^{1-R} - 1)} + O(\frac{1}{m}). \tag{37}$$

$\square$

We shall now consider a special case for which we can derive a slightly better lower bound, viz. a time-varying convolutional code of rate $R = 1/2$ and memory m in which $G_0(j) = (11)$ for $j = 0, 1, 2, \ldots$ and each digit in each of the matrices $G_i(j)$ for $1 \leq i \leq m$ and $j = 0, 1, 2, \ldots$ is chosen independently with probability $1/2$. This code has column distance $d_0 = 2$. We use 2^{i+1} as an upper bound on the number of incorrect paths of length $m + i + 1, i \geq 0$. The probability for each of these paths which start with the symbols 11 on the first branch is $2^{-(m+i)2}$ and there are $\binom{(m+i)2}{j}$ ways of choosing exactly j ones among the last $(m + i)2$ code symbols.

As a counterpart to inequality (28) we have

$$P[d_\infty < d] < \sum_{j=0}^{d-3} \sum_{i=0}^{\infty} 2^{i+1} \binom{(m+i)2}{j} 2^{-(m+i)2}. \tag{38}$$

Using the substitution (29) and following the same steps as in the previous derivation we obtain

$$P[d_\infty < d] < \frac{2^{1-m}}{(2^R - 1)(2^{1-R} - 1)^{d-2}}. \tag{39}$$

Thus, if

$$\frac{2^{-m}}{(2^{1-R} - 1)^d} \leq \frac{2^R - 1}{2(2^{1-R} - 1)^2} = \frac{1}{2(\sqrt{2} - 1)}, \tag{40}$$

there exists at least one code of rate $R = 1/2$ and memory m with $d_\infty \geq d$. Since the right side of (40) is greater than one we obtain

$$d \leq \frac{m}{- \log_2(\sqrt{2} - 1)} = 0.79m, \tag{41}$$

and we have proved

Theorem 8 (Lower bound—special case): There exists a binary, time-varying convolutional code of rate $R = 1/2$ and memory m with a free distance satisfying the inequality

$$\frac{d_\infty}{m} \geq 0.79. \tag{42}$$

$\square$

6 Lower bound on distance profile

In this section we shall derive a new lower bound on the distance profile for the ensemble of fixed convolutional codes.

Theorem 9: There exists a fixed, binary convolutional code of rate $R = b/c$ and memory m whose column distances satisfy

$$d_j \geq \rho c(j+1), \tag{43}$$

for $0 \leq j \leq m$, and where ρ is the Gilbert-Varshamov parameter, i.e. the solution of

$$h(\rho) = 1 - R, \tag{44}$$

where $h(\rho) = -\rho \log_2 \rho - (1 - \rho) \log_2(1 - \rho)$ is the binary entropi function. $\square$

Before proving this theorem we show in Fig. 6 the optimum distance profile [CeJ88] together with the lower bound (43) for a rate $R = 1/2$ ($\rho = 0.11$) fixed, binary convolutional code.

Proof: Let $d_{0,l}, 0 \leq l < 2^b$, denote the weights of the branches with $i_0 \neq 0$ stemming from the root. From Theorem 2 we have

$$P[d_{0,l} = k] = \binom{c}{k}\left(\frac{1}{2}\right)^c, \tag{45}$$

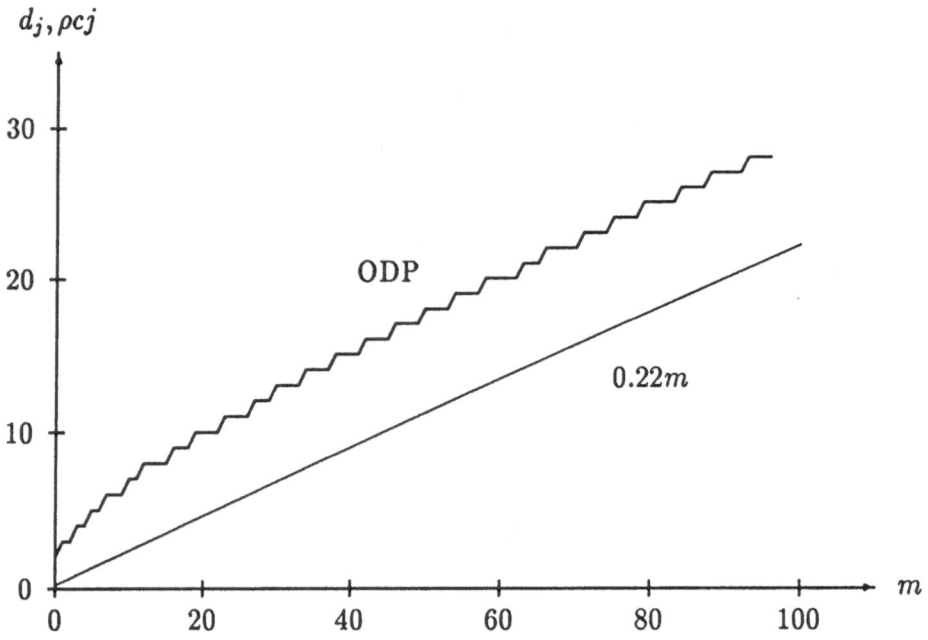

Figure 6: ODP and its lower bound for rate $R = 1/2$ and $0 \leq m \leq 96$.

for $0 \le k \le c$ and $0 \le l < 2^b$. Consider all paths stemming from the lth initial branch with $i_0 \ne 0$ for $0 \le l < 2^b$, i.e. these paths begin at depth 1 and not at the root!

Now let us introduce the following random walk:

$$S_j = \sum_{i=1}^{j} \sum_{s=1}^{c} Y_{is}, \tag{46}$$

where $Y_{is} = \alpha$ if the sth symbol on the $(i+1)$th branch is 1 and $Y_{is} = \beta$ if it is 0. Let w_j be the weight of the corresponding path of length j branches. Then equation (46) can be rewritten as

$$S_j = w_j \alpha + (jc - w_j)\beta. \tag{47}$$

Furthermore, since $d_j = k + w_j$ we notice that if we choose $\alpha = 1$ and $\beta = 0$, then $k + S_j$ should stay above the straight line in Fig. 6 for $1 \le j \le m$ in order to guarantee the existence of a code satisfying our bound. Since it is more convenient to analyze a situation with an absorbing barrier parallel to the x-axis we choose

$$\begin{cases} \alpha &= 1 - \rho \\ \beta &= -\rho \end{cases} \tag{48}$$

and introduce an absorbing barrier at $c\rho - k$. By inserting (48) into (47) we obtain

$$\begin{aligned} S_j &= w_j(1-\rho) - (jc - w_j)\rho \\ &= w_j - \rho c j. \end{aligned} \tag{49}$$

Thus the random walk stays above the barrier, i.e.

$$S_j > c\rho - k, \tag{50}$$

if and only if

$$d_j = k + w_j > \rho c(j+1). \tag{51}$$

To estimate the probability that the random walk is absorbed we introduce a random variable ξ_{ji} such that $\xi_{ji} = 1$, if the random walk for a path leading to the ith node at depth j will cross the barrier for the first time at depth j, and $\xi_{ji} = 0$ otherwise.

The average of the random variable ξ_{ji} is equal to the probability that the random walk S_j drops below the barrier $c\rho - k$ for the first time at depth j, i.e.

$$E[\xi_{ji}] = P[S_n > c\rho - k, 1 \le n < j, \& S_j = v, v \le c\rho - k]. \tag{52}$$

Summing (52) over $1 \leq j \leq m$ and $1 \leq i \leq 2^{bj}$ we get

$$\sum_{j=1}^{m} \sum_{i=1}^{2^{bj}} E[\xi_{ji}] < \sum_{j=1}^{\infty} E[\xi_{ji}] 2^{bj}. \tag{53}$$

Using the notation from Appendix (76) we have

$$E[\xi_{ji}] = \sum_{v \leq c\rho - k} f_{0,j}(c\rho - k, v). \tag{54}$$

Hence, the right side of inequality (53) can be rewritten as

$$\sum_{j=1}^{\infty} \sum_{v \leq c\rho - k} f_{0,j}(c\rho - k, v)(2^{-b})^{-j}. \tag{55}$$

Then we introduce (c.f. Wald's identity):

$$\sum_{j=1}^{\infty} \sum_{v \leq c\rho - k} f_{0,j}(c\rho - k, v) 2^{rv} (g(r))^{-j}, \tag{56}$$

where

$$g(r) = E[2^{rZ_i}] \tag{57}$$

is the moment-generating function of the random variable

$$Z_i = \sum_{s=1}^{c} Y_{is}. \tag{58}$$

Choose

$$r_0 = \log_2 \frac{\rho}{1 - \rho}. \tag{59}$$

Then we have

$$
\begin{aligned}
g(r_0) &= \left(\frac{1}{2} 2^{r_0 \alpha} + \frac{1}{2} 2^{r_0 \beta} \right)^c \\
&= \left(\frac{1}{2} 2^{(1-\rho) \log \frac{\rho}{1-\rho}} + \frac{1}{2} 2^{-\rho \log \frac{\rho}{1-\rho}} \right)^c \\
&= \left(\frac{1}{2} \left(\left(\frac{\rho}{1-\rho} \right)^{1-\rho} + \left(\frac{\rho}{1-\rho} \right)^{-\rho} \right) \right)^c \\
&= \left(\frac{1}{2} \rho^{-\rho} (1-\rho)^{1-\rho} \right)^c \\
&= 2^{(-1+h(\rho))c} = 2^{-Rc} = 2^{-b}.
\end{aligned} \tag{60}
$$

Combining (56) and (60) we get

$$\sum_{j=1}^{\infty} \sum_{v \le c\rho - k} f_{0,j}(c\rho - k, v)2^{r_0 v}2^{bj}$$

$$\ge 2^{r_0(c\rho - k)} \sum_{j=1}^{\infty} \sum_{v \le c\rho - k} f_{0,j}(c\rho - k, v)2^{bj}$$

$$= 2^{r_0(c\rho - k)} \sum_{j=1}^{\infty} E[\xi_{ji}]2^{bj}, \tag{61}$$

where we have used the fact that $r_0 < 0$.

Since

$$g'(r)\mid_{r=r_0} = c\left(\frac{1}{2}2^{r_0\alpha} + \frac{1}{2}2^{r_0\beta}\right)^{c-1}\left(\frac{\alpha}{2}2^{r_0\alpha} + \frac{\beta}{2}2^{r_0\beta}\right)\ln 2$$

$$= c\left(\frac{1}{2}2^{r_0\alpha} + \frac{1}{2}2^{r_0\beta}\right)^{c-1}\left(\frac{1-\rho}{2}2^{(1-\rho)\log\frac{\rho}{1-\rho}} - \frac{\rho}{2}2^{-\rho\log\frac{\rho}{1-\rho}}\right)\ln 2$$

$$= c\left(\frac{1}{2}2^{r_0\alpha} + \frac{1}{2}2^{r_0\beta}\right)^{c-1}\left(\frac{\rho}{1-\rho}\right)^{-\rho}\left(\frac{\rho}{2} - \frac{\rho}{2}\right)\ln 2 = 0$$

we use (91) and upper bound the probability that the random walks stemming from a node l at depth 1 with column distance $d_{0,l} = k$ are absorbed by $2^{-r_0(c\rho-k)}$.

Finally, summing over all nodes at depth 1 with $i_0 \ne 0$ we get

$$(2^b - 1)\sum_{k=0}^{c}\binom{c}{k}\left(\frac{1}{2}\right)^c 2^{-r_0(c\rho-k)}$$

$$= (2^{b-c} - 2^{-c})2^{c\rho\log\frac{1-\rho}{\rho}}\sum_{k=0}^{c}\binom{c}{k}2^{-k\log\frac{1-\rho}{\rho}}$$

$$= (2^{(R-1)c} - 2^{-c})2^{c\rho\log\frac{1-\rho}{\rho}}\left(1 + 2^{-\log\frac{1-\rho}{\rho}}\right)^c$$

$$= (2^{-h(\rho)c} - 2^{-c})\left((1-\rho)^{-(1-\rho)}\rho^{-\rho}\right)^c = (2^{-h(\rho)c} - 2^{-c})2^{ch(\rho)}$$

$$= 1 - 2^{-c(1-h(\rho))} = 1 - 2^{-b} < 1, \tag{62}$$

for $0 < \rho < 1/2$.

Since the probability of absorbtion is strictly less than 1, there exists a fixed convolutional code with a distance profile satisfying the bound and the proof is complete. □

7 Discussion

In this paper we have given an overview of the most important distance measures for convolutional codes as well as upper and lower bounds on the free distance.

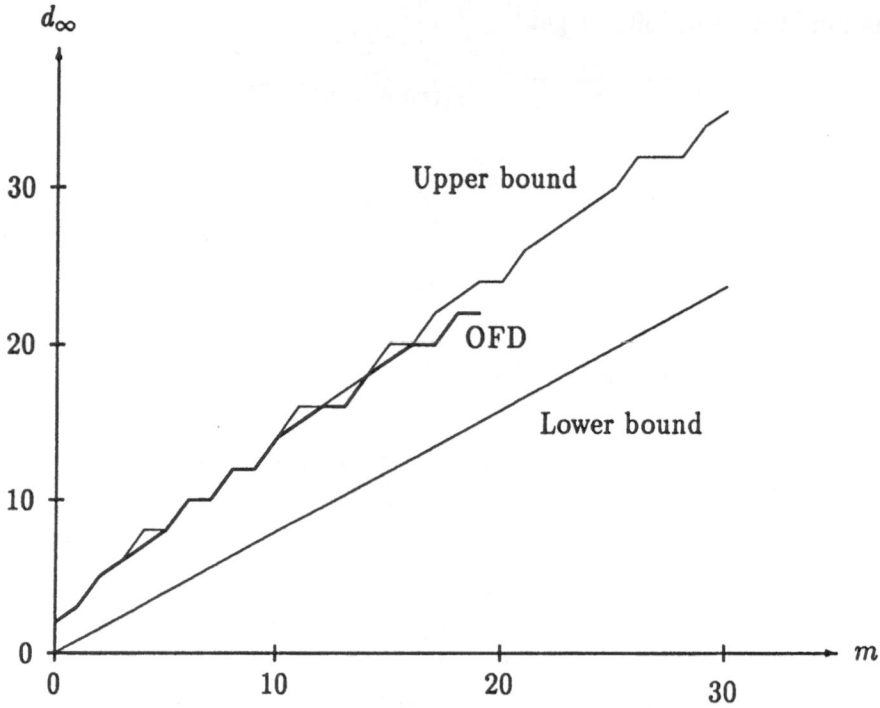

Figure 6: Upper and lower bounds on the free distance and the free distance for OFD codes.

For rate $R = 1/2$, binary convolutional codes we have calculated Heller's upper bound for memories $m \leq 30$. Using Griesmer's bound we improved this bound for some values of m by one or two. The result is shown in Fig. 6 and compared with the free distance for optimum free distance (OFD) *fixed* convolutional codes [CeJ88]. The upper bound is surprisingly tight! For comparison we also show the lower bound for *time-varying* convolutional codes. Finally, we have given a proof based on Wald's identity of a new lower bound on the distance profile for fixed convolutional codes.

Appendix. Wald's identity

Let a random walk be represented by the sequence of random variables $S_0 = 0, S_1, S_2, \ldots$, where

$$S_j = \sum_{i=1}^{j} Z_i, \tag{63}$$

for $j \geq 1$, where the Z_i's are independent, identically distributed random variables. We assume that the average $E[Z_i] > 0$ and are interested in bounding the probability that the random walk drops below a barrier at $u < 0$, $P[S_{\min} \le u]$, where $S_{\min} = \min_j S_j$. (Notice that if $E[Z_i] \le 0$, then $P[S_{\min} \le u] = 1$.) We also assume that the random variable Z_i takes on at least one negative value with a probability > 0, since otherwise $P[S_{\min} \le u] = 0$.

Let

$$g(r) = E[2^{rZ_i}] \qquad (64)$$

be the moment generating function of Z_i.

Example:

Let

$$Z_i = \sum_{s=1}^{c} Y_{is}, \qquad (65)$$

where the Y_{is}'s are independent, identically distributed random variables,

$$Y_{is} = \begin{cases} \alpha > 0, & \text{with probability } p \\ \beta < 0, & \text{with probability } 1 - p, \end{cases} \qquad (66)$$

and

$$P[Z_i = k\alpha + (c - k)\beta] = \binom{c}{k} p^k (1 - p)^{c-k}. \qquad (67)$$

Since the Y_{is}'s are independent and identically distributed we have

$$g(r) = E[2^{rZ_i}] = \prod_{l=1}^{c} E[2^{rY_{il}}]$$

$$= \left(p2^{r\alpha} + (1 - p)2^{r\beta}\right)^c. \qquad (68)$$

Furthermore,

$$g(0) = 1 \qquad (69)$$

and

$$\begin{aligned} g'(0) &= c(p2^{r\alpha} + (1-p)2^{r\beta})^{c-1}(p\alpha 2^{r\alpha} + (1-p)\beta 2^{r\beta})\ln 2 \mid_{r=0} \\ &= c(p\alpha + (1-p)\beta)\ln 2 = cE[Y_{is}]\ln 2 \\ &= E[Z_i]\ln 2. \end{aligned} \qquad (70)$$

From (70) it follows that a positive drift of the random walk, i.e. $E[Z_i] > 0$, corresponds to a positive derivative of the moment generating function $g(r)$ at $r = 0$. $\qquad \square$

Now we shall change the probability assignment of the random walk in such a way that the new random walk will have a negative drift and hence will be absorbed with probability one!

We introduce the "tilted" probability assignment

$$q_{Z,r}(z) = \frac{f_Z(z)2^{rz}}{g(r)}, \tag{71}$$

where $f_Z(z)$ is the probability assignment for the original random walk. We notice that

$$q_{Z,r}(z) \geq 0, \text{ all } z \tag{72}$$

and

$$\begin{aligned}
\sum_z q_{Z,r}(z) &= \frac{1}{g(r)} \sum_z f_Z(z)2^{rz} \\
&= \frac{1}{g(r)} E[2^{rz}] = 1.
\end{aligned} \tag{73}$$

Let us introduce the corresponding random walk $S_{0,r} = 0, S_{1,r}, S_{2,r} \ldots$, where

$$S_{j,r} = \sum_{i=1}^{j} Z_{i,r}, \tag{74}$$

for $j \geq 1$, where the $Z_{i,r}$'s are independent, identically distributed random variables.

We find that

$$\begin{aligned}
E[Z_{i,r}] &= \sum_z q_{Z,r}(z)z \\
&= \frac{1}{g(r)} \sum_z f_Z(z)z2^{rz} \\
&= \frac{1}{g(r)\ln 2} \frac{d}{dr} \sum_z f_Z(z)2^{rz} \\
&= \frac{1}{g(r)\ln 2} \frac{dg(r)}{dr}.
\end{aligned} \tag{75}$$

Thus by choosing $r = r_0$ such that $g'(r_0) < 0$ we see that $E[Z_{i,r_0}] < 0$ and our tilted random walk has a negative drift.

A typical moment generating function $g(r)$ is shown in Fig. A.1:

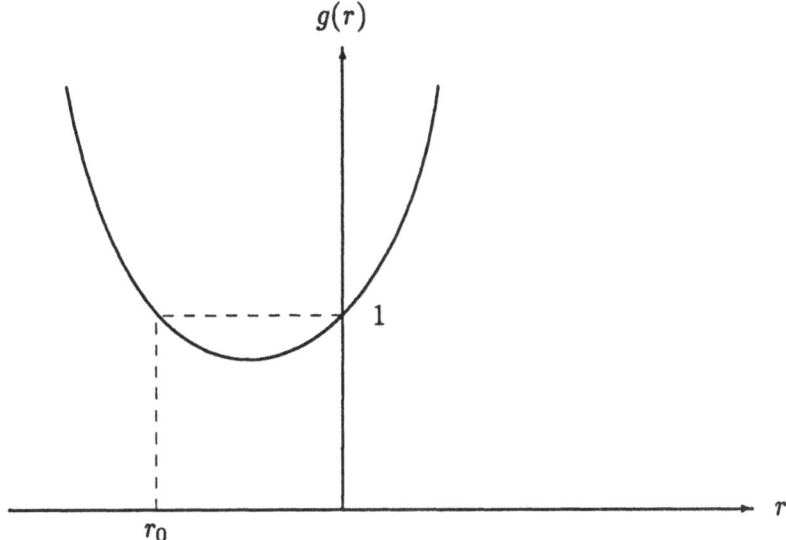

Figure: A.1. Typical $g(r) = E[2^{rZ_i}]$, where $E[Z_i] > 0$.

Since $g'(0) > 0$ we have a positive drift for $r = 0$, $E[Z_i] > 0$, but for $r = r_0$ it follows from $g'(r_0) < 0$ that the drift is negative, i.e. $E[Z_{i,r_0}] < 0$.

In order to estimate $P[S_{\min} \le u]$ we study the tilted random walk. Let $f_{r,t}(u, v)$ denote the probability that it is absorbed by a barrier at $u < 0$ at time t, i.e. for $v \le u$ we have

$$f_{r,t}(u, v) \stackrel{\text{def}}{=} P[S_{n,r} > u, 1 \le n < t, \& S_{t,r} = v]. \tag{76}$$

We notice that

$$\sum_{t=1}^{\infty} \sum_{v \le u} f_{r,t}(u, v) \tag{77}$$

is the probability that the tilted random walk ever achieves a value less than or equal to u. Suppose that for $r = r_0$ the drift is negative, then it follows from the law of large numbers that the tilted random walk will eventually achieve a value less than or equal to u with probability 1, i.e.

$$\sum_{t=1}^{\infty} \sum_{v \le u} f_{r_0,t}(u, v) = 1. \tag{78}$$

From (71) it follows that

$$\prod_{i=1}^{t} P[Z_{i,r_0} = a_i] = \prod_{i=1}^{t} P[Z_i = a_i] 2^{r_0 a_i} g(r_0)^{-1}. \tag{79}$$

Hence we have

$$f_{r_0,t}(u,v) = f_{0,t}(u,v)2^{r_0v}g(r_0)^{-t}, \tag{80}$$

where $\sum_{i=1}^{t} a_i = v$. Combining (78) and (80) we obtain

$$\sum_{t=1}^{\infty}\sum_{v\leq u} f_{0,t}(u,v)2^{r_0v}g(r_0)^{-t} = 1, \tag{81}$$

which is known as *Wald's identity* for a random walk with an absorbing barrier at $u < 0$.

Assuming that $r_0 < 0$ it follows from (81) that

$$2^{r_0u}\sum_{t=1}^{\infty}\sum_{v\leq u} f_{0,t}(u,v)g(r_0)^{-t} \leq 1 \tag{82}$$

or that

$$\sum_{t=1}^{\infty}\sum_{v\leq u} f_{0,t}(u,v)g(r_0)^{-t} \leq 2^{-r_0u}. \tag{83}$$

We can now use (83) to upper bound the probability that the random walk $S_0 = 0, S_1, S_2, \ldots$ drops below a barrier at $u < 0$, i.e. $P[S_{\min} \leq u]$. First we notice that

$$P[S_{\min} \leq u] = \sum_{t=1}^{\infty}\sum_{v\leq u} f_{0,t}(u,v). \tag{84}$$

Let $r_0 < 0$ be a root of the equation

$$g(r) = 1 \tag{85}$$

(see Fig. A.1), i.e. $g(r_0) = 1$. Then from (83) and (84) we have

$$P[S_{\min} \leq u] \leq 2^{-r_0u}. \tag{86}$$

In a more general case we would like to upper bound

$$\sum_{t=1}^{\infty}\sum_{v\leq u} f_{0,t}(u,v)2^{bt}. \tag{87}$$

Then we choose $r = r_0 < 0$ such that

$$g(r_0) = 2^{-b}. \tag{88}$$

Assume that $g'(r_0) < 0$. Thus the tilted random walk has a negative drift and the conditions for equation (78) are satisfied. Hence we can use Wald's identity (81) to obtain

$$\sum_{t=1}^{\infty} \sum_{v \leq u} f_{0,t}(u,v) 2^{r_0 v} 2^{bt} = 1. \tag{89}$$

Since $r_0 < 0$ it follows from (89) that

$$2^{r_0 u} \sum_{t=1}^{\infty} \sum_{v \leq u} f_{0,t}(u,v) 2^{bt} \leq 1 \tag{90}$$

or that

$$\sum_{t=1}^{\infty} \sum_{v \leq u} f_{0,t}(u,v) 2^{bt} \leq 2^{-r_0 u}. \tag{91}$$

Notice that (91) is still valid if $g'(r_0) = 0$ for $r_0 < 0$. Then we have to replace b in (88) by $b - \varepsilon$, where $\varepsilon > 0$. Let $r_0(\varepsilon)$ be the root of

$$g(r) = 2^{-b+\varepsilon}, \tag{92}$$

then $r_0(\varepsilon) < r_0$ and $g'(r_0(\varepsilon)) < 0$. Hence we use Wald's identity and obtain

$$\sum_{t=1}^{\infty} \sum_{v \leq u} f_{0,t}(u,v) 2^{(b-\varepsilon)t} \leq 2^{-r_0(\varepsilon)u}. \tag{93}$$

When $\varepsilon \to 0$ (93) will approach (91).

References

[CeJ88] Cedervall, M. and Johannesson, R. (1988), A FAST algorithm for computing distance spectrum of convolutional codes. Submitted to *IEEE Trans. Inform. Theory*. May 1988.

[ChC76] Chevillat, P. R. and Costello, D. J., Jr. (1976), Distance and computing in sequential decoding. *IEEE Trans. Commun.*, COM-24:440–447.

[Cos69] Costello, D. J., Jr. (1969), A construction technique for random-error-correcting convolutional codes. *IEEE Trans. Inform. Theory*, IT-19:631–636.

[Cos74] Costello, D. J., Jr. (1974), Free distance bounds for convolutional codes. *IEEE Trans. Inform. Theory*, IT-20:356–365.

[For67]   Forney, G. D., Jr. (1967), *Review of random tree codes* (NASA Ames. Res. Cen., Contract NAS2-3637, NASA CR 73176, Final Rep.;Appx A). See also Forney, G. D., Jr. (1974), Convolutional codes II. Maximum-likelihood decoding and convolutional codes III: Sequential decoding. *Inform Contr.*, 25:222-297.

[For70]   Forney, G. D., Jr. (1970), Convolutional codes I: Algebraic structure. *IEEE Trans. Inform. Theory*, IT-16:720-738.

[Gol67]   Golomb, S. W. (1967), *Shift Register Sequences*, Holden-Day, San Fransisco, 1967. Revised ed., Aegean Park Press, Laguna Hills, Cal., 1982.

[Gri60]   Griesmer, J. H. (1960), A bound for error-correcting codes. *IBM J. Res. Develop.*, 4:532–542.

[Hel68]   Heller, J. A. (1968), Short constraint length convolutional codes. Jet Propulsion Lab., California Inst. Technol., Pasadena, Space Programs Summary 37-54, 3:171–177.

[Joh75]   Johannesson, R. (1975), Robustly optimal rate one-half binary convolutional codes. *IEEE Trans. Inform. Theory*, IT-21:464–468.

[LaM70]   Layland, J. and McEliece, R. (1970), An upper bound on the free distance of a tree code. Jet Propulsion Lab., California Inst. Technol., Pasadena, Space Programs Summery 37–62, 3:63–64.

[MaC71]   Massey, J. L. and Costello, D. J., Jr. (1971), Nonsystematic convolutional codes for sequential decoding in space applications. *IEEE Trans. Commun. Technol.*, COM-19:806–813.

[Mas75]   Massey, J. L. (1975), Error bounds for tree codes, trellis codes, and convolutional codes with encoding and decoding procedures, in G. Longo (ed.), *Coding and Complexity—CISM Courses and Lectures No. 216*, Springer Verlag, Wien.

[MaS68]   Massey, J. L. and Sain, M. K. (1968), Inverses of linear sequential circuits. *IEEE Trans. Comput.*, C-17:330–337.

[Vit71]   Viterbi, A. J. (1971), Convolutional codes and their performance in communication systems. *IEEE Trans. Commun. Technol.*, COM-19:751–772.

COMBINED CHANNEL CODING AND CONSTANT AMPLITUDE CONTINUOUS PHASE MODULATION

Carl-Erik W. Sundberg

AT&T Bell Laboratories, Signal Processing Research Department
600 Mountain Avenue, Murray Hill, NJ 07974, USA

ABSTRACT

This paper consists of an overview of recent results on combined constant amplitude continuous phase modulation (CPM) and convolutional channel codes. We also compare the best coded CPM systems to trellis-coded amplitude/phase modulations. The ideal additive white Gaussian channel and systems with perfect carrier phase and symbol time synchronization are considered in this paper. Signals which have a constant amplitude at all times are sometimes needed when highly nonlinear amplifiers are used. Memory is introduced into the constant amplitude signals by means of continuous phase at all times, controlled intersymbol interference to further smooth the phase changes and combinations of CPM and convolutional codes. Analytical tools for calculating the power spectrum and upper bounds on the error probability are given. Results for the special case of Continuous Phase Frequency Shift Keying (CPFSK) with up to 256 levels are considered. Some new results on bounds on more advanced coded systems of higher complexity where the coding is performed jointly over several consecutive channel symbols are also given. Current research topics such as trellis-coded modulation for fading channels, and practical implementations and applications are also briefly discussed.

1. INTRODUCTION

For many of today's communication channels, bandwidth is a limited resource. Due to increased traffic there is a need to transmit at higher information bit rates relative to the channel bandwidth. Examples are voice-band channels, terrestrial digital radio channels, satellite links and mobile radio channels. Modulation schemes that conserve bandwidth are therefore of considerable interest. We know from communication theory that bandwidth efficiency can be achieved at the expense of increased transmitted power. However, in many systems power efficiency is also of great importance. Recently, at the expense of complexity in the form of signal processing, jointly power and bandwidth efficient constructive methods of combining digital modulation and channel coding have emerged. These systems yield improved combined power and bandwidth efficiency over traditional coded and uncoded modulations. Good results are only obtained when the channel coding and the digital modulation units are jointly optimized. Spectral efficiency is often measured in b bits/sec per Hz or alternatively as the 99% power-in-band bandwidth $2B$ Hz. (We will also call this the 99% bandwidth.) Power efficiency is often measured in required E_b/N_0 on a Gaussian channel, where E_b is the average energy per information bit and N_0 is the one sided spectral density of the Gaussian white noise. Shannon theory indicates that present uncoded modulation systems typically operate at about 10 dB higher signal-to-noise ratio than required at capacity for the additive white Gaussian channel and an error probability of 10^{-6}. With recent combined coding and modulation schemes, improvements of 3-6 dB in E_b/N_0 over uncoded modulations can be achieved at high channel signal-to-noise ratios without increasing the bandwidth. In an uncoded modulation system, the receiver makes symbol-by-symbol decisions each symbol time. In combined coding and modulation, the receiver is typically a soft decision Viterbi receiver which makes sequence decisions over several symbol intervals.

Continuous Phase Modulation (CPM) [1], [2] is a unified description of a large family of digital modulation systems which have exactly constant amplitude at all times including moments in time when data symbols change. Such methods are of interest when nonlinear transmitter amplifiers and repeaters are used. Typically, in CPM the information-carrying phase varies smoothly with time for any information sequence. Spectral efficiency in terms of low spectral side lobes and narrow main lobe is achieved by smoothing the information-carrying phase, not the amplitude. Thus the transmitted signal contains memory due to continuous phase and sometimes controlled intersymbol interference. Further memory can be introduced by combining CPM with convolutional codes.

We will also briefly discuss combined coding and modulation using multilevel amplitude modulation (AM) or signal sets with combined amplitude and phase modulation, AMPM. These systems are generally not of constant-amplitude type. Spectral efficiency is achieved either through pre-modulation pulse-shaping or post-modulation filtering. The original trellis-coded modulation work was done for one dimensional (multilevel) AM, two dimensional AM or combined AM and PM and 8-level PSK, [3], [4]. By expanding the signal set from $2^{(m-1)}$ to 2^m points and applying a code of rate $(m-1)/m$, the uncoded signal set with $2^{(m-1)}$ points and the coded set have the same bandwidth, while the coded system is 3-6 dB more power efficient. In more recent work on coded signal sets in more than 2 dimensions [4], some further improvements in gains have been achieved. Recent high speed modems use signal design in up to 16 dimensions. These modems operate at 19.2 kb/s at a bandwidth of about 3 kHz.

Throughout this paper we will primarily use the normalized 99% power-in-band bandwidth $2BT_b$ as a measure of spectral efficiency. The quantity T_b is the information bit time. The data rate is $R_b = 1/T_b$ bits/sec. The 99% bandwidth is a more precise definition than bits/sec/Hz for nonlinear modulation methods like CPM. For power efficiency, we will primarily use the free Euclidean distance as a measure. This will immediately give the required relative signal-to-noise ratio compared to a reference system at low error probability values. This difference is often referred to as the coding/modulation gain in dB. For the purpose of comparing binary and multilevel systems, all signal-to-noise ratio values will be expressed in energy per information bit E_b. For the same purpose, the Euclidean distance values will be given in normalized form. For a few systems, we will give the bit error probability or upper bound it.

Unless otherwise stated, we assume throughout this chapter that an ideal additive white Gaussian channel is used. The class of systems which are of interest to us in this paper are operating under jointly power-limited and bandwidth-limited conditions. Furthermore we will only consider ideal coherent detection with perfect carrier phase and symbol timing synchronization. The memory which is introduced in the signal is used by the receiver by means of a maximum likelihood sequence detector.

In subsection 1.1 we will give some details about a few selected coded modulations for the purpose of giving the reader a few samples of things to come. In subsection 1.2 the class of continuous phase modulation signals is defined. In Section 2 we will summarize results for efficient combined convolutional codes and CPM. This is followed by Section 3, where a brief overview of trellis coded amplitude modulation is given and where we also compare AM and CPM. This book chapter is terminated by a discussion and conclusion section where some new results on CPM are mentioned as well as a brief discussion is given of research topics in combined coding and modulation.

1.1 Examples of Coded Systems

As an introduction to the type of material we will present in this chapter, we will give some important parameters of a few selected coded and uncoded modulation systems. These systems are discussed in some detail in the following sections of this chapter. Table 1 gives estimates of power efficiency, bandwidth and complexity of six different systems. The power efficiency is measured in required E_b/N_0 to achieve a bit error probability of 10^{-6}. The bandwidth $2BT_b$ is the normalized 99% power-in-band bandwidth where we assume that all systems transmit at the same data rate $1/T_b$. Finally, the relative complexity is measured in the number of states required in an optimum maximum likelihood Viterbi detector. System 1 is Minimum Shift Keying (MSK), which is a simple CPM system which can be optimally detected by means of a symbol-by-symbol detector. Systems 2 to 4 are combinations of multilevel Continuous Phase Frequency Shift Keying (CPFSK) modulation and convolutional codes. Note that CPFSK is also a special case of CPM (and MSK is a special case of binary CPFSK). The codes have the rate R (number of information bits over the number of output bits) specified in the table and the code memory is $v = 4$ giving a number of code states of 16. The CPFSK modulations have 8, 16 and 32 levels respectively. The quantity h is the modulation index, which will be defined in Section 1.2. It controls the maximum phase change over a symbol interval. From the power and bandwidth values in the table, we can see that significant improvements can be obtained compared to MSK. Thus we can see that exactly constant amplitude modulation methods with about half the MSK bandwidth can be constructed without loss in error probability performance. Gains in E_b/N_0 of several dB's can be achieved without bandwidth expansion.

	SYSTEM	h	v	Required E_b/N_0	Bandwidth	Complexity
1	MSK	1/2		10.5 dB	1.18	1
2	$R = 2/3$, 8-CPFSK	1/10	4	9.5 dB	0.65	320
3	$R = 3/4$, 16-CPFSK	2/15	4	6.0 dB	0.95	240
4	$R = 4/5$, 32-CPFSK	2/15	4	5.0 dB	1.25	240
5	4-PSK			10.5 dB	0.65	1
6	$R = 2/3$, 8-PSK		4	7.0 dB	0.65	16

Table 1. Performance and complexity of some selected uncoded and coded systems.

For comparison, we have also given data for Nyquist filtered uncoded four phase shift keying (4-PSK or QPSK) and trellis-coded 8-PSK using a combination of rate 2/3 convolutional coding and 8-PSK. Although 4-PSK and 8-PSK systems sometimes are referred to as constant amplitude systems, the signal amplitude varies a lot in between sampling instants. Thus, in practice, there is a strong component of amplitude modulation. Systems 5 and 6 are more bandwidth efficient and simpler than systems 1 to 4 at the expense of variations of the signal amplitude. It is also interesting to compare the systems in Table 1 to the Shannon Capacity bound. Thus, for a system to operate error free at 2 bits per second per Hz, an E_b/N_0 of 2 dB is required. This corresponds to an ideal bandwidth $2B T_b$ of 0.5 in Table 1. For a system at an ideal bandwidth of 1.0, an E_b/N_0 of at least 0 dB is required to obtain error free transmission. The trellis-coded rate 2/3 8-PSK system can be improved by about 2 dB, as we will see later. Thus, by means of trellis-coding we have reduced the gap to channel capacity from about 10 dB to about 5 dB. The reason that the number of states in the Viterbi detector is larger for systems 2 to 4 than for system 6 (although they have the same code memory) is that the CPM systems also utilize the memory in the continuous phase. It is interesting to observe in Table 1, that the complexity increase both in the direction of more power efficient systems and more bandwidth efficient systems than MSK.

In the rest of this book chapter we will give the details about the systems in Table 1. We will also give guidelines for how the performance numbers can be improved. First, the parameters of CPM are defined.

1.2 Definition of Continuous Phase Modulation

First we will describe continuous phase modulation. A large class of constant amplitude continuous phase modulation (CPM) schemes is defined by:

$$s(t) = s(t, \alpha) = \sqrt{\frac{2E}{T}} \cos(2\pi f_0 t + \phi(t, \alpha)) \tag{1}$$

where the transmitted information is contained in the phase:

$$\phi(t, \alpha) = 2\pi h \sum_{i=-\infty}^{\infty} \alpha_i q(t - iT) \tag{2}$$

with $q(t) = \int_{-\infty}^{t} g(\tau) d\tau$. Normally the function $g(t)$ is a smooth pulse shape (in instantaneous frequency) over a finite time interval $0 \leq t \leq LT$ and zero outside. Thus L is the length of the pulse (per unit T) and T is the symbol time. E is the energy per symbol, f_0 is the carrier frequency and h is the modulation index. The M-ary data symbols α_i take values $\pm 1, \pm 3 \cdots \pm(M-1)$. The number M is normally a power of 2. Note that $E = E_b \cdot \log_2(M)$ and $T = T_b \cdot \log_2(M)$. From the definition of the above class of constant amplitude modulation schemes we observe that the pulse $g(t)$ is defined in instantaneous frequency and its integral $q(t)$ is the phase response. The shape of $g(t)$ determines the smoothness of the transmitted information carrying phase. The rate of change of the phase is proportional to the parameter h, which is normally referred to as the modulation index. Above we have normalized the pulse shape $g(t)$ in such a way that $\int_{-\infty}^{\infty} g(t) dt$ is 1/2. This means that for schemes with positive pulses of finite length, the maximum phase-change over any symbol interval is $(M-1)h\pi$.

Thus, by choosing different pulses $g(t)$ and varying the parameters h and M, a great variety of uncoded CPM schemes can be obtained. Some of the more popular pulse shapes during recent years [1] are listed either explicitly or implicitly in [1], [2], such as Continuous Phase Frequency Shift Keying (CPFSK), Tamed Frequency Modulation(TFM), Generalized TFM (GTFM), Gaussian MSK (GMSK), Duobinary FSK (2REC), Raised Cosine (LRC) and Spectrally Raised Cosine (LSRC). We use the notation LRC for a raised cosine pulse of length L symbol intervals. Thus, 3RC is a raised cosine pulse of length $3T$. The rectangular pulse of length L is denoted LREC. The pulse of length $1T$, that is 1REC, is often referred to as CPFSK in the literature. The 2REC pulse with length $2T$ is also called duobinary. Figure 1 gives the framework for the constant amplitude systems considered in this section. The sequence of binary information symbols u is transmitted over a coded CPM system. For the uncoded CPM system, the input to the M-level mapper is also u. We will later introduce the convolutional code. An example of a mapper for 8 levels is the natural binary scheme [1]. For the uncoded systems we have also used Gray mappers [1]. The additive white Gaussian noise is $n(t)$ and the received signal is $r(t) = s(t, \alpha) + n(t)$. The demodulated sequence of information bits after the Viterbi detector is û.

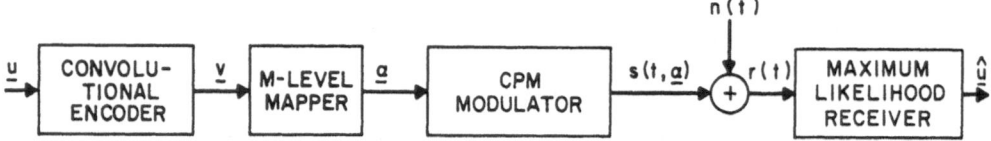

Figure 1. The coded modulation scheme, channel and receiver.

It is convenient to view CPM as a generalization of Minimum Shift Keying (MSK) which is obtained as a special case of the phase modulation signals defined in (1) by selecting the pulse 1REC ($L=1$) and using binary ($M=2$) data with $h = 1/2$. A detailed description of MSK is given in [1]. Minimum Shift Keying is an important special case which can also be viewed as binary Frequency Shift Keying and also as offset quadrature linear modulation in the inphase and quadrature components. Thus, MSK is a special case of nonlinear modulation which in effect is linear. This perhaps explains the popularity of MSK over its truly nonlinear CPM relatives [1].

The CPM signal can in general be viewed either as phase modulation or as frequency modulation. For understanding the optimum coherent receiver, it is advantageous to view the signal as phase modulation. However, it is important to recognize that "modulation" takes place both in frequency and in phase. Memory is introduced into the uncoded CPM signal by means of the continuous phase. Each information carrying phase function $\phi(t, \alpha)$

is continuous at all times for all combinations of data symbols. Even when the data symbols are uncorrelated, further memory can be built into the uncoded CPM signal by choosing a $g(t)$ pulse with $L > 1$. These schemes have overlapping pulse shaping and are sometimes called partial response techniques. CPM signals with $L = 1$ are referred to as full response schemes. In this case all the memory is in the continuous phase.

Although the CPM signals in (1) are in principle conceivable for any value of the modulation index h, a key to the development of practical maximum likelihood detectors and digitally implemented transmitters is to consider CPM schemes with rational values of h. For $h = 2\ell/p$ where ℓ and p have no common factors, the phase $\phi(t, \alpha)$ during interval $nT \leq t \leq (n+1)T$ can be written:

$$\phi(t, \alpha) = 2\pi h \sum_{i=n-L+1}^{n} \alpha_i q(t - iT) + \theta_n = \theta(t, \alpha) + \theta_n \tag{3}$$

where $\theta_n = h\pi \sum_{i=-\infty}^{n-L} \alpha_i$ modulo 2π can assume only p different values and is referred to as the phase state. $\theta(t, \alpha)$ is completely described by $(\alpha_{n-1}, \ldots, \alpha_{n-L+1})$ and is called the correlative state. Thus the total number of states that is needed at most to describe the signal in (1) is $S = pM^{(L-1)}$ where a state is defined as the vector $(\theta_n, \alpha_{n-1}, \alpha_{n-2}, \ldots, \alpha_{n-L+1})$. For a full response system the number of states is $S = p$.

There are a large number of methods available in the literature for calculating the power spectrum of CPM [1]. Computer simulations can of course also be employed to estimate the power spectra. Several analytical methods are described in [1] for most values of M, h and pulse shapes $g(t)$ with pulse length L [1]. For large values of L, a fast and flexible method using auto correlation function calculations is given. For some very special cases, closed form solutions exist. Bandwidth does not have a unique definition. To illustrate the spectral efficiency of the signal for example, we can define bandwidth as the frequency band around the carrier frequency f_0 containing 99 percent of the signal power. There are also methods in the literature for calculating the power spectra of combined coding and CPM (see [1] and comments below). Throughout this section we will use the methods for uncoded CPM and scale the bandwidth appropriately with the rate of the code. This yields an estimate of the bandwidth for coded CPM, which is very accurate for good convolutional code/CPM combinations, [1],[2].

Figure 2 shows the (one-sided) power spectral density of some CPM schemes. The spectral sidelobes are determined by the smoothness of the pulse $g(t)$. For most practical applications, the raised cosine pulses (LRC) probably have sufficient smoothness. Figure 2 also illustrates that a longer pulse $g(t)$ yields narrower power spectra for fixed h and M. For further examples, see [1] and [2]. Increasing h yield wider power spectra.

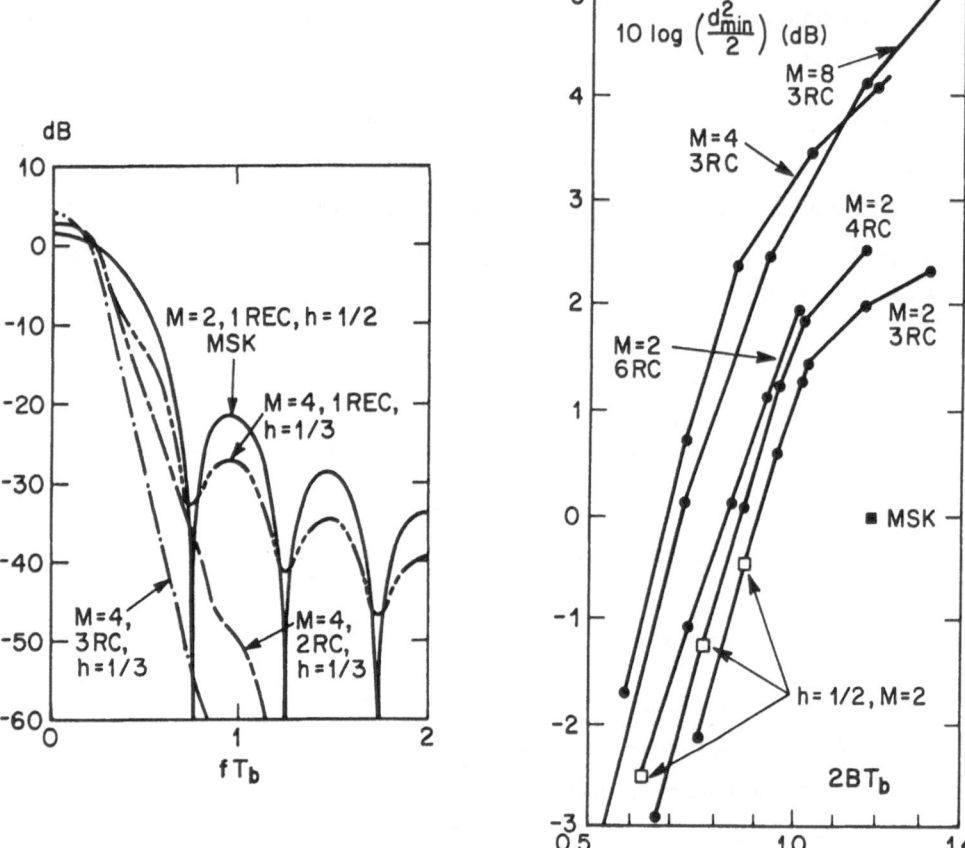

Figure 2. Average power spectra for selected   Figure 3. Power-bandwidth tradeoff for
CPM schemes.   CPM schemes using RC-pulses.

In Figure 3 we consider the joint power and spectral efficiency of uncoded CPM signals. It can be shown that coherent maximum likelihood sequence detection can be performed for all CPM schemes that can be described by the finite state (and trellis) description given above. Although the structure of the optimum ideal coherent receiver for CPM is known [1], it is in general difficult to evaluate its bit error probability performance. Simulations are required for low channel signal-to-noise ratios. The most convenient and useful parameter for describing the error probability of CPM schemes with maximum likelihood detection (Viterbi detection) is the minimum squared Euclidean distance between all possible pairs of signals

$$D_{\min}^2 = d_{\min}^2 \cdot 2E_b = \min \left\{ 2E_b \log_2(M) \frac{1}{T} \int_0^{NT} [1 - \cos[\phi(t, \alpha) - \phi(t, \beta)]] \, dt \right\} \qquad (4)$$

where E_b is the signal energy per bit given by $E_b \log_2(M) = E$ and where NT is the receiver observation interval length in symbol intervals. The minimization is carried out over all $\alpha \neq \beta$ [1]. When N is sufficiently large, the free distance (the largest obtainable minimum distance) is reached. As pointed out above, we would like to compare systems with different values of M. For that purpose, we prefer to work with the signal-to-noise ratio E_b/N_0. Thus, the parameter of interest to us is the normalized squared Euclidean distance $d_{\min}^2 = D_{\min}^2/2E_b$. In most books and papers (but not in all), the normalization is with respect to $2E_b$. The value of $d_{\min}^2$ for MSK is 2. For ideal coherent transmission over an additive white Gaussian channel, the bit error probability for high signal-to-noise ratios, E_b/N_0, is, to a good approximation given by (5) where C is a constant.

$$P_b \approx C e^{-d_{\min}^2 \cdot \frac{E_b}{N_0}}. \qquad (5)$$

Efficient algorithms exist for computing the minimum distance for different $g(t)$, L, h and M [1]. A comparison of CPM schemes is given in the scatter plot in Figure 3 where each point represent a CPM system with its 99 percent power bandwidth $2BT_b$ ($T_b = T/\log_2(M)$) and the signal-to-noise ratio difference relative to MSK in dB for high E_b/N_0. The coding/modulation gain relative to MSK in dB is defined as $10 \log_{10}(d_{\min}^2/2)$. From [1] and from (5) it follows that this gain corresponds to the difference in E_b/N_0 in dB at the same error probability for high channel signal-to-noise ratio values E_b/N_0. This quantity is often referred to as the asymptotic coding/modulation gain. Thus, on the vertical axis in Figure 3 we give $10 \log_{10}(d_{\min}^2/2)$ since the normalized squared minimum Euclidean distance for MSK is 2. In the scatter plot schemes on the same vertical line (for example, through the MSK point) have the same bandwidth for equal data rates. Schemes on the same horizontal line have the same error probability for high signal-to-noise ratios (the same normalized free Euclidean distance). Thus, it is evident that larger L and larger M yield more efficient systems for jointly power and bandwidth limited channels. Not surprisingly, the system complexity increases in the same directions. We have marked some binary ($M = 2$) $h = 1/2$ schemes, some binary ($M = 2$) and quaternary ($M = 4$) 3RC schemes in Figure 3.

Above we have briefly summarized the performance of uncoded CPM from [1], [2]. We will now discuss the further improvements obtained by introducing more memory by means of convolutional coding.

2. EFFICIENT COMBINED CONVOLUTIONAL CODES AND CPM

2.1 Constructive Rate $(n-1)/n$ Codes for 2^n-level CPFSK

The general system description in [1] and Figure 1 still holds, but in this section we first study a restricted class of high rate k/n convolutional encoders [5]-[7]. These have $k = n - 1$, with $n = 3, 4, 5, 6$. Furthermore, the first $(n - 2)$ binary data symbols in each block of k are not coded at all. Thus for the restricted class of codes the coding used is a conventional (in general nonsystematic) rate 1/2 convolutional encoder which encodes the data symbol sequence $\ldots, u_k^{(-1)}, u_k^{(0)}, u_k^{(1)}, \ldots$ to produce the two coded symbol sequences $\ldots, v_{n-1}^{(-1)}, v_{n-1}^{(0)}, v_{n-1}^{(1)}, \ldots$ and $\ldots, v_n^{(-1)}, v_n^{(0)}, v_n^{(1)}, \ldots$. Formally this can be written as

$$v_m^{(\ell)} = u_m^{(\ell)} \qquad\qquad m = 1, 2, \ldots, n-2$$

$$v_m^{(\ell)} = \sum_{i=0}^{\nu} g_{m,k}(i) u_k^{(\ell-i)}, \qquad \ell = \ldots, -1, 0, 1, \ldots, \qquad m = n-1, n \qquad (6)$$

where $g_{m,k}(i) \in \{0, 1\}$ and represents the connection between $u_k^{(\ell-i)}$ and $v_m^{(\ell)}$. The summation is modulo 2 and ν is the number of delay elements in the encoder. The state x of the encoder has ν binary components and there are 2^ν states. Since the encoder described above has $(n - 1)$ inputs and n outputs, the rate $R_c = k/n$ of the encoder is $(n-1)/n$ data symbols/coded symbol. We also earlier used the simpler notation R for the code rate.

The coded sequence v is the input to a mapper which associates levels in an M-ary alphabet with blocks of coded symbols. The output from the mapper is a sequence α of channel symbols $\alpha_\ell \in \{\pm 1, \pm 3, \ldots, \pm(M-1)\}$ where M is a power of 2. The mapper has rate $R_m = \log_2(M)$ coded symbols/channel symbol, which is fixed at n. The overall rate R_s is therefore $R_s = R_c R_m = (n-1)$ binary data symbols/channel symbol. The (natural) M level mapping rule used is defined by

$$\alpha_\ell = \sum_{i=0}^{n-1} v_{\ell n + i} \cdot 2^{(n-i)} - M + \ell, \quad \ell = 0, \pm 1, \pm 2, \cdots \qquad (7)$$

The symbols $v_{\ell n}$ and $v_{(\ell+1)n-\ell}$ are referred to as the most significant bit (MSB) and the least significant bit (LSB), respectively. The channel encoding in this section can be viewed as a conventional rate 1/2 convolutional encoder on the least significant symbol in the mapping rule. For further details, see [1]. In this section it is assumed that the frequency pulse of the CPM modulator is given by

$$g(t) = \begin{cases} [1 - \cos(\pi t/\epsilon T)]/4T, & 0 < t < \epsilon T \\ 1/2T, & \epsilon T \leq t \leq T \\ \{1 + \cos[\pi(t-T)/\epsilon T]\}/4T, & T < t < (1+\epsilon)T \\ 0, & \text{otherwise} \end{cases} \tag{8}$$

with $0 \leq \epsilon \leq 1$. When $\epsilon = 0$ and $\epsilon = 1$ we have the CPFSK scheme (1REC) and the 2RC scheme, respectively. Thus (8) defines a class of frequency pulses in between CPFSK and 2RC, and a certain degree of smoothness and partial response can be controlled by the parameter ϵ.

For CPFSK modulation ($\epsilon = 0$) we will search for the optimal rate 1/2 noncatastrophic [1] convolutional encoders such that, for given v and h, d_{min}^2 is maximized. (A catastrophic convolutional code is such that a finite number of channel errors can cause an infinite number of decoding errors [8], [9].) Some of the optimal encoders are then used together with a frequency pulse with a small ϵ, $\epsilon > 0$. The idea is to create a modulation scheme which has roughly the same d_{min}^2 as the corresponding coded CPFSK scheme but with considerably smaller side lobes in the spectrum because of the smoothing and partial response.

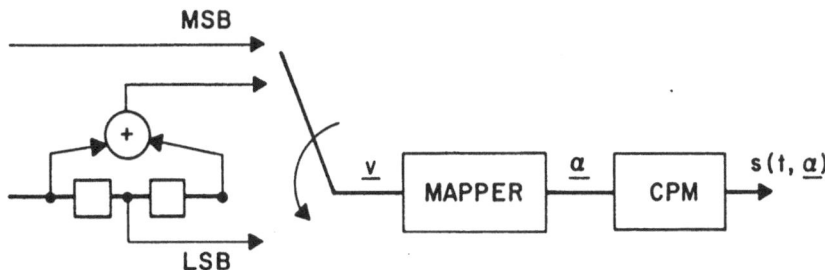

Figure 4. The rate 2/3 (5,2)-encoder combined with eight-level CPM modulation.

As an example consider the coded modulation scheme in Figure 4. This scheme consists of the rate 2/3 optimal (5,2)-encoder combined with eight-level CPFSK modulation. The convolutional encoder is specified by means of its connection polynomial in octal form. The upper one is given first. Now assume that the start state of the encoder is $x = (0, 0)$. If the two information bit sequences $u_\alpha = (01, 10, 00, 01)$ and $u_\beta = (10, 01, 10, 01)$ are fed into the encoder the corresponding phase functions are $\phi(t, \alpha)$ and $\phi(t, \beta)$. These phase functions are linearly increasing or linearly decreasing depending on the corresponding channel symbols; see the example in Figure 5. It is easy to show that the two data symbol sequences u_α and u_β given above generate a phase and state merger after four channel symbol intervals [1]. Thus an upper bound on d_{min}^2 for the specific coded CPFSK scheme given in Figure 4 can be calculated following methods in [1]. This upper bound d_{β}^2 is

$$d_{\min}^2 \leq d_{\beta}^2 = 2 \left[4 - \frac{2}{3} \frac{\sin(2\pi h)}{2\pi h} - \frac{10}{3} \frac{\sin(4\pi h)}{4\pi h} \right]. \tag{9}$$

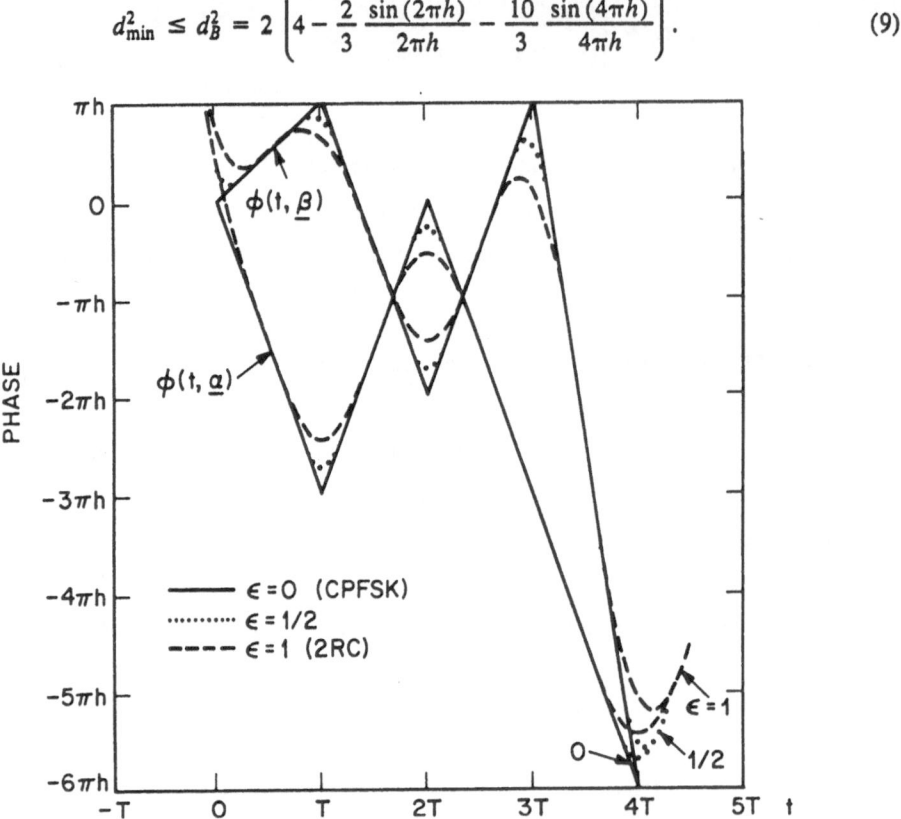

Figure 5.   Examples of phase and state mergers for the modulation scheme in Figure 4 with channel symbol sequences $\alpha = (-3, 3, -3, -3)$ and $\beta = (1, -3, 3, -7)$.

Let us now employ a frequency pulse $g(t)$ with the parameter $\epsilon \neq 0$. Using the sequences u_α and u_β we again generate phase and state mergers for these cases. We add a common channel symbol at the end of the two channel symbol sequences α and β generated by u_α and u_β, and because of the partial response we must also specify a common start-symbol. Figure 5 shows the corresponding phase functions $\phi(t, \alpha)$ and $\phi(t, \beta)$ when $\epsilon = 1/2$ and $\epsilon = 1$; the start symbol is -7 and the added channel symbol is 3. Note that the phase functions are shown with different delays for different values of the parameter ϵ. This makes Figure 5 easier to read. It is seen that a phase and state merger occurs after five channel symbol intervals, and an upper bound on $d_{\min}^2$ for the specific coded CPM scheme $(0 < \epsilon \leq 1)$ given in Figure 4 can be calculated from this merger. By increasing the parameter ϵ the phase function $\phi(t, \alpha)$ becomes more smooth, and it is this smoothing that reduces the sidelobes in the spectrum. This is shown in Figure 6.

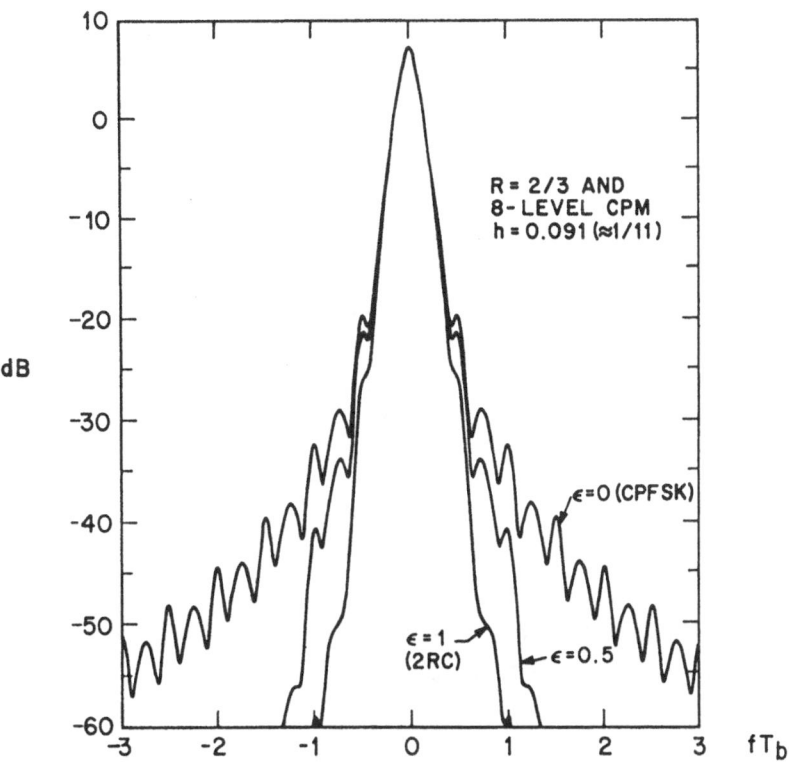

Figure 6. Examples of estimated coded power spectra in dB for different ε-values.

In [1] it is shown that the Euclidean distance between any two signals $s(t, \alpha)$ and $s(t, \beta)$ is a function only of $\gamma = \beta - \alpha$. By making the encoder input number $n-1$ always 0 (i.e., the second (lower) input to the encoder in Figure 4), it is possible to generate phase and state merges corresponding to the difference sequences $\gamma = (..., 0, 8\ell, -8\ell, 0, ...)$, $\ell = 1, 2, ..., 2^{n-2} - 1$. This means that the rate 1/2 code is always yielding an all zero output. These phase and state mergers are independent of the actual rate 1/2 convolutional encoder used. Taking the minimum of these $(2^{n-2} - 1)$ upper bounds and assuming CPFSK modulation yields the encoder-independent upper bound (see [1])

$$d^2_{min} \leq d^2_B = 2(n-1) \min_{1 \leq k \leq K} \left(1 - \frac{\sin(k8\pi h)}{k8\pi h}\right) \tag{10}$$

where $K = 2^{n-2} - 1$. These upper bounds are often tight. Convolutional encoders together with 2^n-level CPFSK modulation reaching this bound are given in [1]. In particular we are interested in encoders with small v values. As we increase n the number of encoder-independent weak modulation indices is also increased, owing to the distance properties of the encoded 2^{n-2} level CPFSK scheme. Weak modulation indices are such rational h-values where (10) cannot be reached. If the reduction in distance is very large, we sometimes call these h-values catastrophic. See [1] for details.

For the case of rate 2/3 convolutional encoders with eight-level CPFSK modulation, the encoder consists of an uncoded MSB and a rate 1/2 code on the remaining bit, as in Figure 4. The optimum (largest d_{min}^2) rate 1/2 convolutional encoders for this case have been found when $1 \leq v \leq 4$. Some of these rate 1/2 encoders can then be used in rate 3/4 and rate 4/5 convolutional encoders combined with 16-level and 32-level CPFSK modulation. The specific modulation indices studied were $h = 2/22, 2/21, \ldots, 2/9, 2/8$. Examples of numerical results for optimum codes are presented in Tables 4 to 6. For further details, see [1]. The encoders in Table 2 are given by means of the connection polynomials in octal form. All the codes are rate 1/2 codes with the upper polynomial given first.

h	$v = 2$		$v = 3$		$v = 4$	
	d_{min}^2	Encoder	d_{min}^2	Encoder	d_{min}^2	Encoder
1/11	1.43	(5,2)	1.84	(15,2)	2.16	(33,4)
1/10	1.71	(5,2)	2.19	(15,2)	2.59	(33,4)
1/9	2.07	(5,2)	2.66	(15,2)	3.14	(33,4)
1/8	2.56	(5,2)	3.28	(15,2)	3.89	(33,4)
1/7	3.22	(5,2)	4.13	(15,2)	4.48	(33,4)
1/6	4.14	(5,2)	4.76	(15,2)	4.82	(25,10)
2/11	3.90	(3,4)	4.82	(11,2)	4.87	(32,1)

Table 2. Optimal encoders for rate 2/3, 8-level CPFSK systems.

Very good schemes of combined rate $(n-1)/n$ convolutional encoders and 2^n level CPFSK modulation, $n = 3, 4, 5$, have been found when the modulation index is in the interval $0 \leq h \leq 1/4$. Examples of good rate 1/2 encoders are the (2,1)-encoder, the (5,2)-encoder, the (15,2)-encoder, and the (33,4)-encoder. When using these four rate 1/2 encoders, the shortest minimum distance mergers at small h are $\gamma = (-4, 6, -2)$, $\gamma = (-4, 6, -6, 4)$, $\gamma = (-4, 8, -6, 6, -4)$, and $\gamma = (-4, 4, 2, 0, -2, -4, 4)$, respectively. These γ sequences can be used to construct code dependent upper bounds on d_{min}^2, [1], [5].

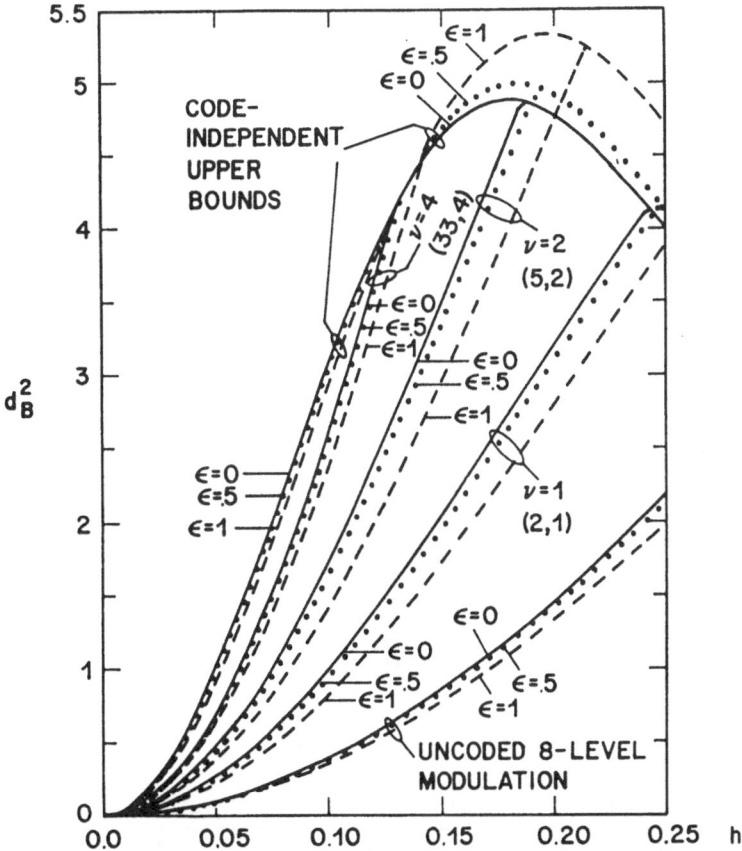

Figure 7. Upper bounds on d_{min}^2 for rate 2/3 convolutionally coded eight-level CPM.

Some of the encoder-dependent upper bounds d_B^2 on d_{min}^2 are shown in Figure 7 for $n = 3$. Also shown are the corresponding encoder-independent upper bound (10) and upper bound on uncoded 8-level CPFSK. Since many of the upper bounds are tight in the CPFSK case, these figures actually show d_{min}^2 for some specific coded CPFSK schemes, [1]. For the same four encoders, we can replace the CPFSK frequency pulse with a smoother one from (8) and look at the effect of the smoothness parameter ϵ on the upper bounds. As shown in Figure 7 the bounds do not change much if $0 \le \epsilon \le 1/2$. We can expect that d_{min}^2 for small ϵ will be similar to the CPFSK case. Figure 8 shows power-bandwidth tradeoffs for some of the constructive coded CPFSK schemes. On the vertical axis is the gain in d_{min}^2 compared to MSK and on the horizontal axis is the estimated normalized double sideband 99% bandwidth. Figure 8 depicts the rate 2/3 encoders given in Table 2 combined with eight-level CPFSK modulation. Specific schemes are shown as rectangles. The coded schemes become more power efficient with increasing ν, until the encoder-independent upper bound is reached.

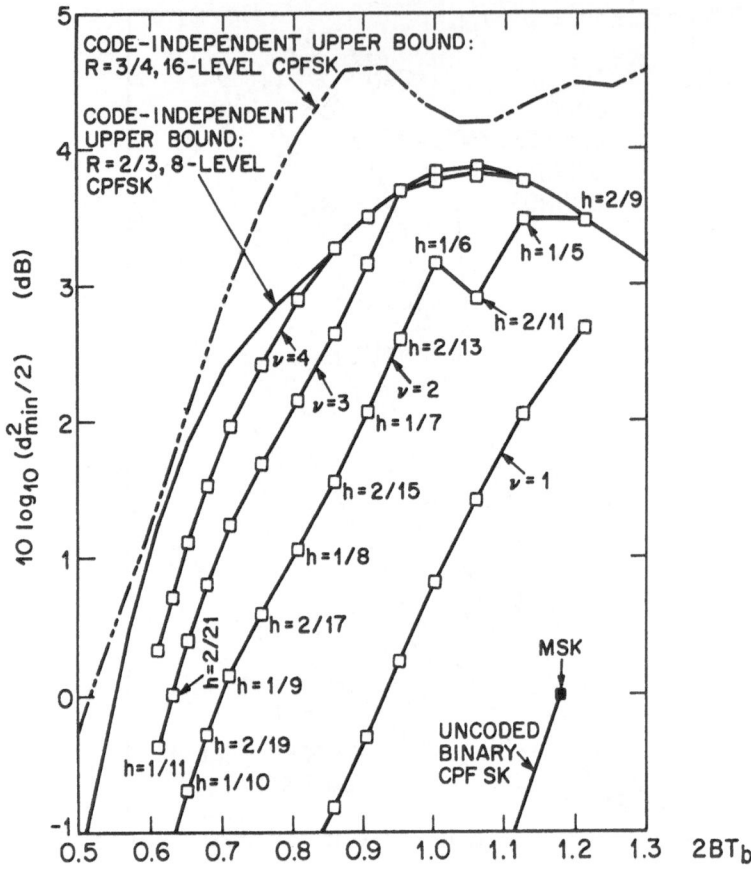

Figure 8. Power-bandwidth tradeoff for some coded 8-level CPFSK schemes.

Figure 8 also considers schemes consisting of the rate 3/4 encoders combined with 16-level CPFSK modulation. These schemes also become more power efficient with increasing v, until the encoder-independent upper bound is reached. Figure 9 considers the best schemes found in [1]. An example of such a scheme is the (33,4)-encoder combined with 16-level CPFSK modulation and $h = 2/17$ with a gain of about 4.2 dB and a bandwidth which is 74% of the MSK bandwidth. Figure 9 also shows some of the best known multi-h CPFSK schemes (denoted A-D). Multi-h CPFSK systems are obtained, when the modulation index h is time-varying over successive symbol intervals in a periodic manner. Multi-h systems are simple examples of coded CPFSK systems. Binary and 8-level two-h systems are inferior to 4-level systems [1]. Table 3 gives the relevant parameters for the best known 4-level two-h schemes. These are, however, clearly outperformed by the coded CPFSK schemes considered here.

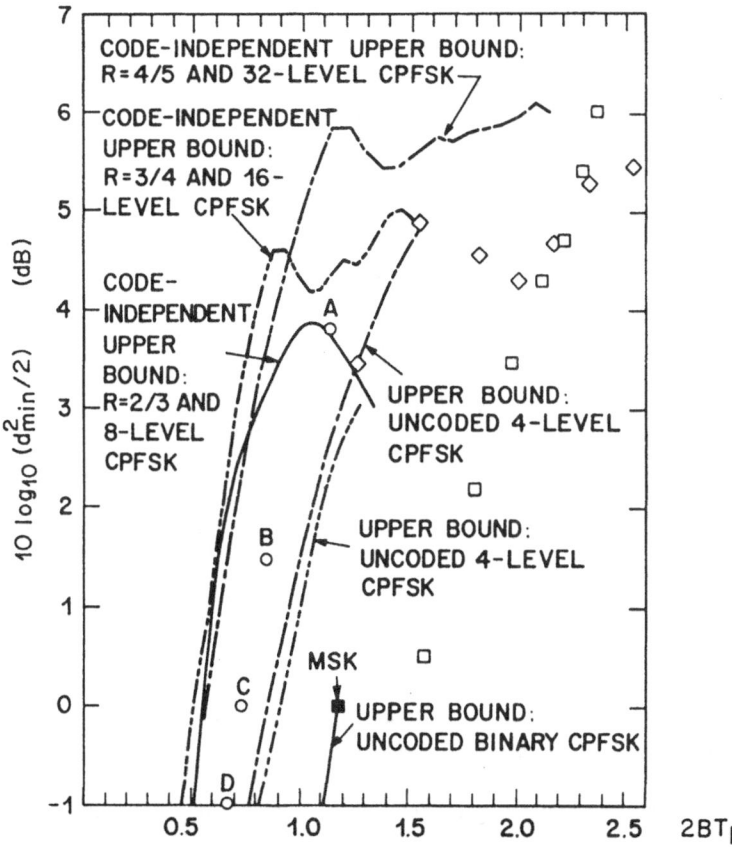

Figure 9. Power-bandwidth tradeoff. The best found $R = 1/2$, binary CPFSK schemes and for the best found $R = 1/2$, 4-level CPFSK schemes are shown for comparison. For multi-h schemes A, B, C, D, see Table 3.

Figure 9 also shows the best rate 1/2 coded binary (□) and quaternary (◇) CPFSK schemes [1] and the multilevel high rate coded CPFSK schemes. Multilevel schemes are clearly better at the price of higher complexity.

Scheme	h_1	h_2	d_{min}^2	$2BT_b$
A	0.40	0.46	4.80	1.13
B	0.25	0.29	2.80	0.84
C	0.25	0.23	2.00	0.73
D	4/16	3/16	1.60	0.67

Table 3. Parameters of 4-level two-h schemes. $2BT_b$ is the 99% bandwidth.

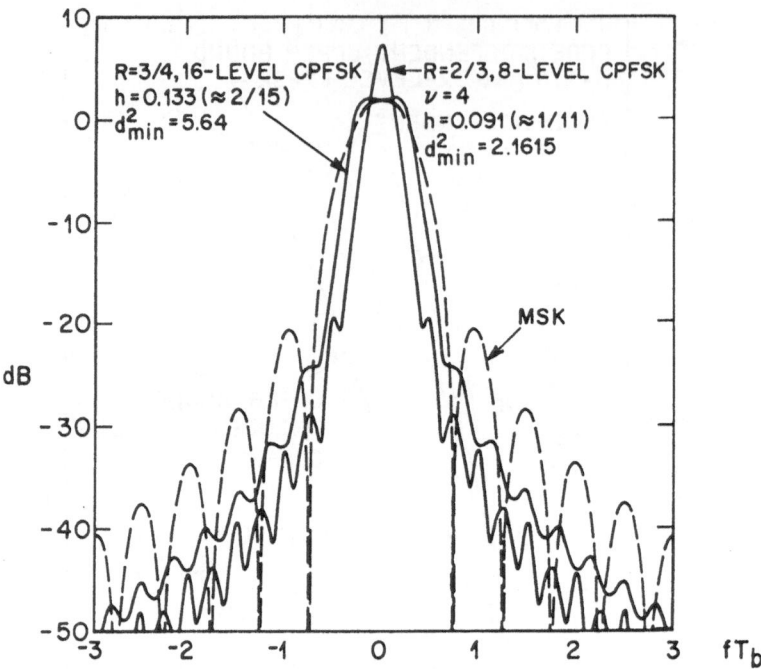

Figure 10. Estimated power spectra for two coded CPFSK schemes. The power spectra for uncoded MSK is given as a reference.

Finally, Figure 10 shows the comparison between the (estimated) power spectra of 2 different coded CPFSK schemes and the calculated power spectrum for minimum shift keying (MSK). One scheme is the $v = 4$, $h = 1/11$, rate 2/3, 8-level scheme with $d^2 = 2.16$. This scheme has the same asymptotic error probability as MSK but only half the 99% bandwidth. Figure 10 also shows the $v = 4$, $h = 2/15$, rate 3/4, 16-level scheme with $d^2_{min} = 5.64$. This system has about the same bandwidth as MSK but a power gain of about 4.5 dB (asymptotically for large signal to noise ratios).

2.2 Free Distance of CPM Systems with Coordinated Coding over Successive

In the previous section we considered constructive coded multilevel CPFSK systems with a code operating over one CPFSK symbol, i.e. the convolutional code rate k/n is such that the number of levels M is related to n by $n = \log_2(M)$. The systems in Section 2.2 are such that $n > \log_2(M)$. The corresponding generalization of trellis-coded AMPM systems is from two dimensional signal sets to multidimensional signal sets. The coded 2^m-level CPFSK schemes considered in this section is shown in Figure 11, [6], [7]. The output from the encoder consists of bA information bits ($1 \leq b \leq m-1$) and cA coded bits ($cA \geq 2$). The $(b+c)A$ bits are partitioned into A subblocks where each subblock has length $m = b + c$ bits. The A

subblocks generate the output from a CPFSK modulator with $M = 2^m$ levels. It is assumed that the natural binary mapping rule is used. Note that the overall system rate is $(mA - 1)/mA$ while the convolutional code rate is $(cA - 1)/cA$.

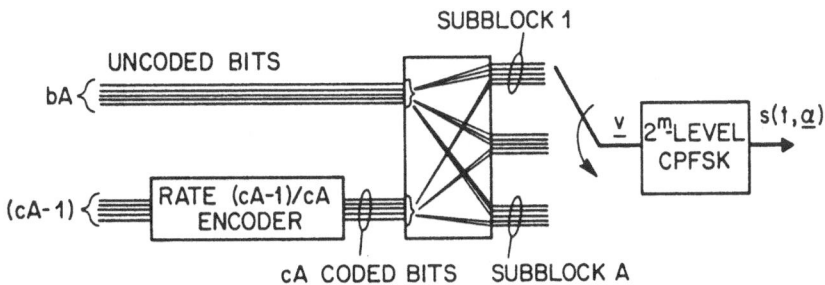

Figure 11. The general coded 2^m-level CPM-scheme with coding over A CPM symbols.

By using the same technique as in [1] it is quite easy to show that the normalized squared minimum Euclidean distance d^2_{min} for the system described above is upper bounded by

$$d^2_{min} \leq d^2_B = 2(m - 1/A) \min_{\ell} \left\{ 1 - \frac{\sin(\pi h \ell 2^{(c+1)})}{\pi h \ell 2^{(c+1)}} \right\} \tag{11}$$

where the minimization is taken over all integers ℓ in the interval $1 \leq \ell \leq (2^b - 1)$. This bound is valid for any code. We are especially interested in small modulation indices h (narrowband schemes), $0 \leq h \leq 2^{-(c+1)}$, for which (11) becomes

$$d^2_{min} \leq d^2_B = 2(m - 1/A) \left(1 - \frac{\sin(\pi h 2^{(c+1)})}{\pi h 2^{(c+1)}} \right). \tag{12}$$

The 99% bandwidth for the modulation scheme in Figure 11 is estimated by

$$2\hat{B}T_b = 2BT_b \cdot \frac{mA}{mA - 1} \tag{13}$$

where $2BT_b$ is the normalized double sided 99% bandwidth for the *uncoded* 2^m-level CPFSK scheme. $2BT_b$ is calculated by numerically integrating the equivalent baseband spectrum $G(\beta)$, $\beta = fT_b$ (see [1] and references therein). The baseband spectrum can be calculated quite fast by the following expression from [7] for noninteger values of h,

$$G(\beta) = \frac{m}{M} \left[S_1(\beta) + \frac{2}{M \cdot N(\beta)} (S_2(\beta)S_3(\beta) + S_4(\beta)S_5(\beta)) \right] \tag{14}$$

where $M = 2^m$ and where $N(\beta) = 1 - 2C \cos(2\pi\beta m) + C^2$, $C = \frac{2}{M} \sum_{n=1}^{M/2} \cos((2n - 1)\pi h)$,

$$S_1(\beta) = \sum_{n=1}^{M} (\sin(\gamma_n)/\gamma_n)^2, \quad S_2(\beta) = \sum_{n=1}^{M} f_1(\beta,n)\sin(\gamma_n)/\gamma_n, \quad S_3(\beta) = \sum_{n=1}^{M} \cos(\pi h n) \sin(\gamma_n)/\gamma_n,$$

$$S_4(\beta) = \sum_{n=1}^{M} f_2(\beta,n) \sin(\gamma_n)/\gamma_n, \ S_5(\beta) = \sum_{n=1}^{M} \sin(\pi h n) \sin(\gamma_n)/\gamma_n \text{ and where}$$

$$f_1(\beta, n) = \cos(2\pi\beta m - \pi h(n-1-M)) - C\cos(\pi h(n-1-M)) \qquad (15)$$

$$f_2(\beta, n) = \sin(2\pi\beta m - \pi h(n-1-M)) + C\sin(\pi h(n-1-M)) \qquad (16)$$

$$\gamma_n = \gamma_n(\beta) = \pi\beta_m - (2n - M - 1)\pi h/2. \qquad (17)$$

When the structure is known (c, A and m in Figure 11) it is now quite straightforward to calculate the upper bound $d_{\hat{B}}^2$ and the estimated bandwidth $2\hat{B}T_b$ for different modulation indices h. As an example see Figure 12 where some results with $m = 4$ (16-level CPFSK) are shown. The corresponding values for the uncoded MSK-scheme (binary CPFSK with $h = 1/2$) are also shown as a reference. It is seen in Figure 12 how the potential asymptotic performance versus bandwidth depends on the choice of parameters c and A when $m = 4$. Compared to the MSK-scheme, large gains in both E_b/N_0 and bandwidth are possible. Specific constructive coded CPFSK schemes with very good power-bandwidth performance for the case $c = 2, A = 1$ and $m = 3, 4$ and 5 can be found in [1] and references therein. Above we have considered some cases for $c = 2$, $A = 1$ and $m = 3$. Compare the results in Figure 12 to those in Figures 8 and 9. Note the potential improvement in both power and bandwidth efficiency.

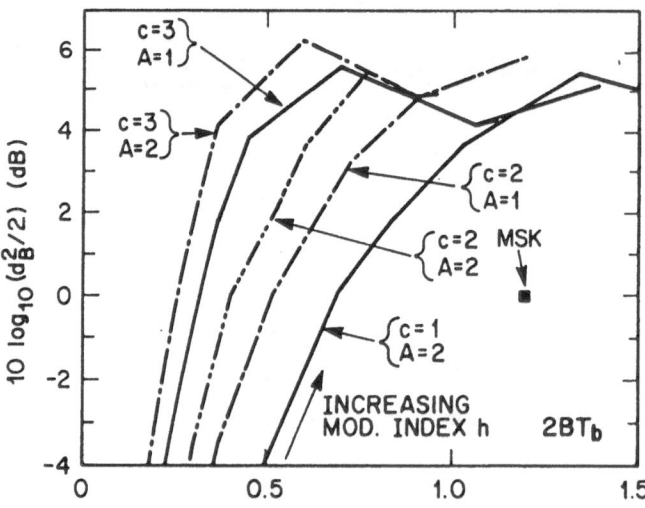

Figure 12. Power-bandwidth tradeoff for the system in Figure 11 for different c and A for $m = 4$.

Table 4 gives some alternative combinations of parameters A, b, c and m. (The set Δ is used in (18).) The performance of these systems in terms of power versus bandwidth is shown in Figures 13 and 14. The results in Figures 12-14 indicate that significantly better systems are possible than those in Section 2.1. Note that for $c > 2$ (see Table 4), we do not know the quality of the code independent upper bound. Further work has to be done to establish the tightness of these bounds and the effectiveness and complexity of constructive codes.

Structure	m	c	b	A	Δ
I	3	1	2	1, 2,...	4, 8, 12
II	3	2	1	1, 2,...	8
III	4	1	3	1, 2,...	4, 8, 12, 16, 20, 24, 28
IV	4	2	2	1, 2,...	8, 16, 24
V	4	3	1	1, 2,...	16

Table 4. Characteristics of structures I-V.

We are particularly interested in modulation schemes operating at less than half the MSK bandwidth. For this purpose, we are investigating 5 structures, denoted by I, II, III, IV and V respectively, see Table 4. Structures I-II use 8-level CPFSK modulation and structures III-V use 16-level CPFSK modulation. The modulation index h is typically in the interval $0.05 \leq h \leq 0.15$. The general encoder structure has been given in Figure 11. Structure I is given with $m = 3$, $c = 1$ and $b = 2$ and it is seen that the overall rate is $(3A - 1)/3A$ where $A = 1$ gives the previously considered systems. It is also seen that the output from this encoder consists of A blocks, where each block has length 3 bits. Furthermore, the first two bits in each block are information bits. The remaining A bits are produced by a rate $(A - 1)/A$ convolutional encoder. For a specific value of A we get a specific class of encoders. Structure II does also consist of a rate of $(3A - 1)/3A$ but now only the first bit in each block of 3 bits is a pure information bit, see Table 4. The remaining $2A$ bits are produced by a rate $(2A - 1)/2A$ convolutional encoder. Structures III-V follow the same ideas. In these cases the overall rate is $(4A - 1)/4A$. It follows from [1] and the above that the normalized squared free Euclidean distance can be upper bounded by

$$d_{min}^2 \leq d_B^2 = 2B \cdot \min_{\delta \in \Delta} \left(1 - \frac{\sin(\pi h \delta)}{\pi h \delta}\right) \tag{18}$$

where $B = (3A - 1)/A$ for Structures I-II and $B = (4A - 1)/A$ for Structures III-V. The set Δ depends on the structure, see Table 4. Figures 13 and 14 show a comparison in terms of both potential power gain (large E_b/N_0) in dB relative to MSK, i.e. $10\log_{10}(d_b^2/2)$, and the estimated 99% bandwidth $2BT_b$. It is seen in Figure 13 that the potential asymptotic performance of Structure II (with $A \geq 2$) is much better than for Structure I. Note however, that we do not know if there exist constructive codes in the different classes which reach the upper bound on d_{min}^2 in this bandwidth region. However, for Structure II with $A = 1$ we know good explicit codes from [1] and above. Figure 14 considers Structures III-V. For Structure IV with $A = 1$ good explicit codes are also reported in [1]. It is seen that the potential asymptotic performance of this class is essentially the same as for Structure I with a very large value on A ($A \to \infty$). It is also seen that the potential asymptotic performance for Structure V with $A \geq 2$ outperforms the other structures in this bandwidth region.

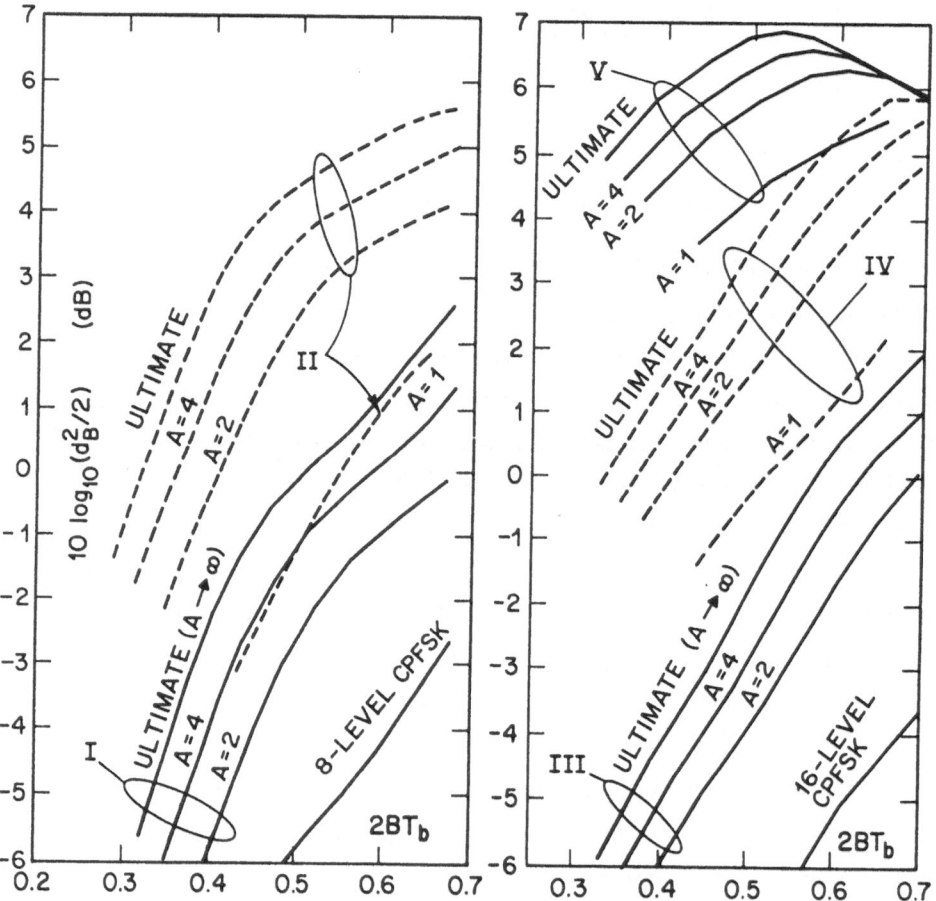

Figure 13. Potential asymptotic performance for structures I-II.

Figure 14. Potential asymptotic performance for structures III-V.

Table 5 shows calculated 99% bandwidth values for *uncoded M-ary CPFSK* with low h. A parameter tradeoff reveals that for $A = 1$ and $c = 2$, $m = 3$ to 4 is close to optimum. For $A = 2$ and $c = 2$, $m = 3$ is optimum. Increasing m does not yield improvements in the low bandwidth (low h) region.

h	M							
	2	4	8	16	32	64	128	256
1/256						0.10	0.142	0.204
0.005				0.04	0.081	0.125	0.163	0.247
1/128							0.234	0.34
0.01	0.0098	0.0231	0.0533	0.101	0.15	0.1905	0.282	0.409
1/64						0.272	0.388	0.59
0.03	0.079	0.138	0.1985	0.2426	0.313	0.437	0.651	1.041
1/32					0.327	0.453	0.675	1.08
0.05	0.183	0.247	0.259	0.325	0.455	0.656	1.011	1.665
1/16				0.407	0.543	0.787	1.235	
1/8	0.548	0.456	0.538	0.677	0.943	1.44		

Table 5. The 99% bandwidth $2BT_b$ for different uncoded M-level CPFSK schemes.

Finally Table 6 shows some selected distance bound values (d_B^2) and 99% bandwidth values for good coded systems. For each system, the d_B^2 value is given and below that, the 99% bandwidth value is given. We can for example see that for fixed c and A, both the distance value and the bandwidth increase. The distance gains for these systems are only obtained by increasing bandwidth.

M	$c = 1, h = 1/4$		$c = 2, h = 1/8$				$c = 3, h = 1/16$			
	$A = 2$		$A = 1$		$A = 2$		$A = 1$		$A = 2$	
8	5	1.07	4	0.81	5	0.65	–	–	–	–
16	7	1.34	6	0.90	7	0.77	6	0.54	7	0.47
32	9	1.91	8	1.18	9	1.05	8	0.68	9	0.60
64	11	–	10	1.73	11	1.57	10	0.94	11	0.86
128	13	–	12	–	13	–	12	1.44	13	1.33

Table 6. Code independent upper bound d_B^2 (left numbers in squares) and bandwidth values (right numbers in squares) for some coded systems. $h = 2^{-(c+1)}$.

2.3 An Upper Bound on the Bit Error Probability of Combined Convolutional Coding and Continuous Phase Modulation

Throughout the discussion above, our performance evaluation of the power efficiency of coded and uncoded CPM has been based on the minimum Euclidean distance. This gives precise insight into the error probability behavior for large channel signal-to-noise ratios E_b/N_0 [1]. However, we would also like to know the error probability also at low and intermediate channel signal-to-noise ratios.

The transfer function technique for evaluating upper bounds on the bit error probability of convolutional codes was introduced in [8] and is described in detail in [9]. It was first applied to conventional linear convolutional codes. This technique was later extended to the more general cases of, e.g. nonlinear trellis codes, in [10]. Also, upper bounds on the symbol error probability for uncoded continuous phase modulation is evaluated in [1]. In [11]-[13], a detailed performance evaluation is presented for some trellis-coded AM and PSK systems. Here we have applied the average transfer function technique to the case of coded continuous phase modulation, [5], [14], [15]. We have derived a general expression for an upper bound on the bit error probability for partial response CPM with finite memory, rational modulation index, an M-ary natural binary mapper and a general convolutional code. An ideal Viterbi detector with infinite path memory is assumed. The upper bound is evaluated numerically for a number of interesting coded multi-level full response CPFSK schemes. Simulation results for a Viterbi detector with finite path memory are also presented at low signal-to-noise ratios and compared to the upper bound. It is shown in [1] and above that CPM systems combined with convolutional codes are jointly more power and bandwidth efficient than uncoded CPM. The previous analysis is based on asymptotic error probability performance for high signal-to-noise ratios given by the minimum Euclidean distance. These conclusions also hold for a range of signal-to-noise ratios when the upper bound on the bit error probability is used for performance evaluation. The gains given by calculations based on the free distance alone *are actually sometimes pessimistic*. An interesting result from the detailed numerical evaluation to follow is that the minimum Euclidean distance error event for some coded CPM schemes only occurs for rather few transmitted signals. This follows from the fact that combined convolutional coding and CPM forms a nonlinear trellis code.

Denote a sequence of information bits by $\mathbf{u} = (\cdots u_{-1}, u_0, u_1, \cdots)$, where $u_j \in \{0, 1\}$ for all j. It is assumed that these bits are generated at a constant rate of $1/T_b$ bits/second. The sequence $\mathbf{u}$ is encoded by a conventional convolutional encoder (k inputs and n outputs), see Figures 1 and 4. The output from this encoder is a sequence of coded binary symbols which is denoted by $\mathbf{v} = (\cdots v_{-1}, v_0, v_1, \cdots)$, where $v_j \in \{0, 1\}$, for all j. It is convenient to partition the sequence $\mathbf{u} = (\cdots , \mathbf{u}_{-1}, \mathbf{u}_0, \mathbf{u}_1, \cdots)$ of information bits into blocks $\mathbf{u}_j$ of length k where $\mathbf{u}_j = (u_{j,1}, u_{j,2}, \ldots, u_{j,k}) = (u_{jk}, u_{jk+1}, \ldots, u_{jk+k-1})$. In a similar way we also partition the sequence $\mathbf{v} = (\cdots , \mathbf{v}_{-1}, \mathbf{v}_0, \mathbf{v}_1, \cdots)$ into blocks $\mathbf{v}_j$ of length n where $\mathbf{v}_j = (v_{j,1}, v_{j,2}, \ldots, v_{j,n}) = (v_{jn}, v_{jn+1}, \ldots, v_{jn+n-1})$. The rate, R_c, of the convolutional encoder is therefore $R_c = k/n$ information bits/coded symbol. The total number of delay elements in the encoder is denoted by ν. Furthermore, the state of the convolutional encoder is denoted by $\mathbf{x} = (x_1, x_2, \ldots, x_\ell, \ldots, x_\nu)$ with $x_\ell \in \{0, 1\}$. Thus, the number of states of the convolutional encoder equals 2^ν. The coded sequence $\mathbf{v}$ is the input to a mapper. This mapper associates levels in an M-ary alphabet with blocks of coded binary symbols. The output from the mapper is a sequence $\boldsymbol{\alpha} = (\cdots , \alpha_{-1}, \alpha_0, \alpha_1, \cdots)$ of channel symbols according to some mapping rule. It is assumed that M is a power of 2 and that $\alpha_j \in \{\pm 1, \pm 3, \ldots, \pm(M-1)\}$, all j. The rate, R_m, of the mapper is therefore $R_m = \log_2(M)$ coded symbols/channel symbol, and the overall rate, R, is $R = R_c R_m$ information bits/channel symbol.

The mapping rules considered will be natural binary (for coded CPM) and Gray M-level (for uncoded CPM) mapping rules will be considered [1]. The combination of the convolutional encoder and the mapping rule can now be described in a convenient way. Let us first define k' and N': k' equals the smallest number of information bits needed to produce an integer number N' of channel symbols [15]. Formally, k' is defined as the smallest positive integer such that (k'/k) equals an integer and such that $(k'/k)n = R_m N'$, where N' is a positive integer. It is seen that each time k' information bits enter the convolutional encoder, the mapper will produce N' channel symbols. The channel symbol sequence $\boldsymbol{\alpha}$ is the input to a continuous phase modulator, which produces the transmitted signal $s(t, \boldsymbol{\alpha})$. It is assumed that $g(t)$ has finite length $\Lambda' T$, i.e. $g(t) \equiv 0$ for $t < 0$ and for $t > \Lambda' T$. Furthermore it is assumed that $q(\Lambda' T) = 1/2$ (as above, see Section 1.2). In general, the pulse length Λ' is allowed to be any real number. In the state description we will use the smallest integer larger than or equal to Λ'. This will be denoted by Λ below. We also use L above, but in the transfer function we will use L for the length of the error event. Let level j correspond to the time $t = jN'T$ and let interval j correspond to the time-interval $jN'T \le t \le (j+1)N'T$. The phase function $\phi(t, \boldsymbol{\alpha})$ can, in interval j, be divided into two parts,

$$\phi(t, \boldsymbol{\alpha}) = 2\pi h q(\Lambda T) \sum_{i=-\infty}^{jN'-\Lambda} \alpha_i + 2\pi h \sum_{i=jN'-\Lambda+1}^{(j+1)N'-1} \alpha_i q(t - iT). \tag{19}$$

The phase state of level j, θ_j, is defined by $\theta_j = 2\pi h q(\Lambda T) \sum\limits_{i=-\infty}^{jN'-\Lambda} \alpha_i$ where it is assumed that

the modulation index $h = 2\ell/p$ (ℓ and p integers with no common divisor) and that $q(\Lambda T) = 1/2$ (if $q(\Lambda T) = 0$ then $\theta_j = 0$ for all j). It is shown in [5] that the number of phase states is $\leq p$. The second term in (19) contains $\Lambda - 1$ partial response channel symbols and the N' channel symbols α_j'. To be able to define the state associated with this term at level j we use the parameter η defined as the smallest non-negative integer such that $\eta N' \geq \Lambda - 1$. The parameter η equals the smallest number of blocks (of length N') needed to incorporate the $\Lambda - 1$ partial response channel symbols. We can now define the state σ_j of the transmitted signal $s(t, \alpha)$ at level j by

$$\sigma_j = \begin{cases} (\theta_j, x_{j-\eta}, u'_{j-\eta}, u'_{j-\eta+1}, \ldots, u'_{j-1}) & , \text{ if } \Lambda \geq 2 \\ (\theta_j, x_j) & , \text{ if } \Lambda = 1 \end{cases} \tag{20}$$

It is seen that if the input to the convolutional encoder, u'_j, is known and if the state of the transmitted signal at level j, σ_j, is known, then the transmitted signal in interval j and the next state σ_{j+1} is known. The transmitted signal can therefore be described by a trellis with S states $S \leq p 2^{(\nu+\eta k')}$. Furthermore, the number of transitions from each state equals $2^{k'}$. The channel is assumed to be an additive white Gaussian noise channel. Thus the signal available for observation is $r(t) = s(t, \alpha) + n(t)$ where $n(t)$ is a white Gaussian random process having zero mean and one-sided power spectral density N_0. The receiver is assumed to be an ideal coherent receiver, which performs maximum likelihood sequence detection (MLSD). The structure of such receivers is described in detail in [1]. Error events are essential in the performance analysis of the receiver. One of the most important characteristics of an error event is its Euclidean distance. The normalized squared Euclidean distance, for an error event, is given by

$$d^2 = \frac{1}{2E_b} \int\limits_0^{\ell N'T} \Big(s(t, \alpha) - s(t, \hat{\alpha}) \Big)^2 dt$$

where $\hat{\alpha}$ corresponds to the decoded channel symbol sequence. In [15] it is shown that the bit error probability P_b is upper bounded by

$$P_b \leq \sum_d C_d Q\Big(\sqrt{d^2 E_b/N_0} \Big) = C_{d_{min}} Q\Big(\sqrt{d^2_{min} E_b/N_0} \Big) + \text{low magnitude terms} \tag{21}$$

where the summation is over all Euclidean distances in the set of all error events. C_d is a constant which is often referred to as the weight associated with distance d. d^2_{min} is the smallest of these distances and is referred to as the normalized squared free Euclidean distance. $Q(x)$ is defined by $Q(x) = \dfrac{1}{\sqrt{2\pi}} \int\limits_{x}^{\infty} e^{-y^2/2}\, dy$. An error event has three

characteristics that influence the upper bound in (21). These characteristics are: the

Euclidean distance d, the length in information bits $\ell k'$ and the number of information bit

errors i. In [5] it is shown that the S states can be divided into 2^ν groups. All states in a

specific group yield the same set of error events, in terms of the characteristics d, $\ell k'$ and i.

Let $a(s_z, i, \ell, d)$ denote the number of error events of length $\ell k'$ information bits, starting

from any state in group number $z(z = 1, 2, \ldots, 2^\nu)$, having the normalized Euclidean

distance d with i information bit errors. Then C_d in (21) can be written as

$$C_d = \frac{2^{-\nu}}{k'} \sum_i \sum_\ell \sum_{z=1}^{2^\nu} a(s_z, i, \ell, d)\, i \left(\frac{1}{2}\right)^{\ell k'}. \tag{22}$$

P_b can now be upper bounded by

$$P_b \leq P_u = \frac{1}{k'}\, Q\left(\sqrt{d_{\min}^2 E_b/N_0}\right) e^{d_{\min}^2 E_b/2N_0} \left. \frac{\partial T(D, L, I)}{\partial I} \right|_{\substack{D = e^{-E_b/2N_0} \\ L = (1/2)^{k'} \\ I = 1}} \tag{23}$$

where the average generating function $T(D, L, I)$ is

$$T(D, L, I) = 2^{-\nu} \sum_{z=1}^{2^\nu} T_z(D, L, I) = 2^{-\nu} \sum_{z=1}^{2^\nu} \sum_i \sum_\ell \sum_d a(s_z, i, \ell, d) D^{d^2} L^\ell I^i. \tag{24}$$

$T_z(D, L, I)$ administrates the set of error events starting from any of the states in group

number z. To find $T_z(D, L, I)$ we use a method that generates the parameters d^2, ℓ and i for

all error events starting from a given state. The method we use is a so called pair-state

(super-state) description [5],[15]. Thus the bit error probability for coded CPM systems with

Viterbi detection can be upper bounded by using the trellis description for the combined

coding and modulation scheme. The resulting bound is obtained from an averaged

generating function. Convolutional codes yields a linear system and no averaging is required

for the conventional upper bounds. A detailed derivation of the upper bound is given in

[5], [14], [15].

The upper bound on the bit error probability in (23) has been calculated for some

interesting coded, as well as uncoded, M-level full response CPFSK schemes. The average

generating function $T(D, L, I)$ in (24) has been evaluated by truncation of the series-

expansion of a matrix, see [5]. The numerical results are given in figures showing the upper

bound P_u as a function of (E_b/N_0). In these figures we also give the function

$Q\left(\sqrt{d_{\min}^2 E_b/N_0}\right)$ for comparison. Thus the figures indicate when the minimum Euclidean

distance events dominates the upper bound. We can also approximately find the value of

$C_{d_{min}}$. As a reference we often use the uncoded binary full response CPFSK scheme with $h = 1/2$, MSK, which has $d_{min}^2 = 2$. Therefore the error probability of MSK, given by the function $Q\left(\sqrt{2E_b/N_0}\right)$, is also given in the figures. Optimum codes have been selected from search results in [1], [5]. Note that the upper bound in (23) is calculated for a Viterbi detector with *infinite* path memory. In practice the detector memory is of course finite. When this memory is chosen sufficiently large, the upper bound for the Viterbi detector with finite memory will approach the upper bound in (23). The selection of the path memory length depends on the distance structure of the coded CPM scheme, see [1].

First consider the upper bound on the bit error probability for uncoded CPFSK modulation. From earlier studies [1] we know that the pair $\mathbf{u} = (1, 0)$ and $\hat{\mathbf{u}} = (0, 1)$ often gives the minimum Euclidean distance. In general it is not easy to find $C_{d_{min}}$. If however, those pairs of sequences which yields the difference sequence $\alpha - \hat{\alpha} = \gamma = (2, -2)$ or $\gamma = (-2, 2)$ are the only existing minimum distance pairs, then

$$C_{d_{min}} \geq \frac{4}{\log_2(M)} \cdot \left(\frac{M-1}{M}\right)^2$$ for uncoded M-level full response schemes. Examples of both 4-level and 8-level CPFSK schemes, which achieves this lower bound on $C_{d_{min}}$, have been given in [1], when the Gray mappers are used. The upper bound on the bit error probability has been evaluated for one of the coded binary CPFSK schemes given in [1]. The code is defined by the connections in octal form, see [1]. It is seen that the asymptotic behavior starts roughly at levels around 10^{-7} depending on the modulation index. It is seen that $C_{d_{min}}$ is much larger when $h = 1/2$ than it is for the other modulation indices. From [1] we know that $h = 1/2$ is especially interesting, since large d_{min}^2-values have been found for this modulation index. The upper bound P_u on the bit error probability has been evaluated for some of the coded 4-level CPFSK schemes given in [1]. Figure 15 shows the results for the (7,2)-encoder. Note in Figure 15 that $C_{d_{min}}$ is much less than 1 for the considered modulation indices. The upper bound on the bit error probability has also been evaluated for some of the coded 8-level CPFSK schemes found in [5]. The rate 2/3 (15,2)-encoder is considered in Figure 16. For the modulation indices in the figure, all minimum distance pairs have length $5T$. Furthermore, for $h = 1/7$ and for $h = 2/13$ there are other pairs with distances very close to d_{min}^2. When $h = 1/7$ there are pairs of length $4T$ with distance 4.18 which is close to $d_{min}^2 = 4.13$. For $h = 2/13$ there are pairs of length $2T$ with distance 4.69 (the encoder-independent upper bound) which is close to $d_{min} = 4.67$. By varying the modulation index in the interval $1/10 \leq h \leq 2/13$ various combinations of power saving and bandwidth saving schemes are obtained.

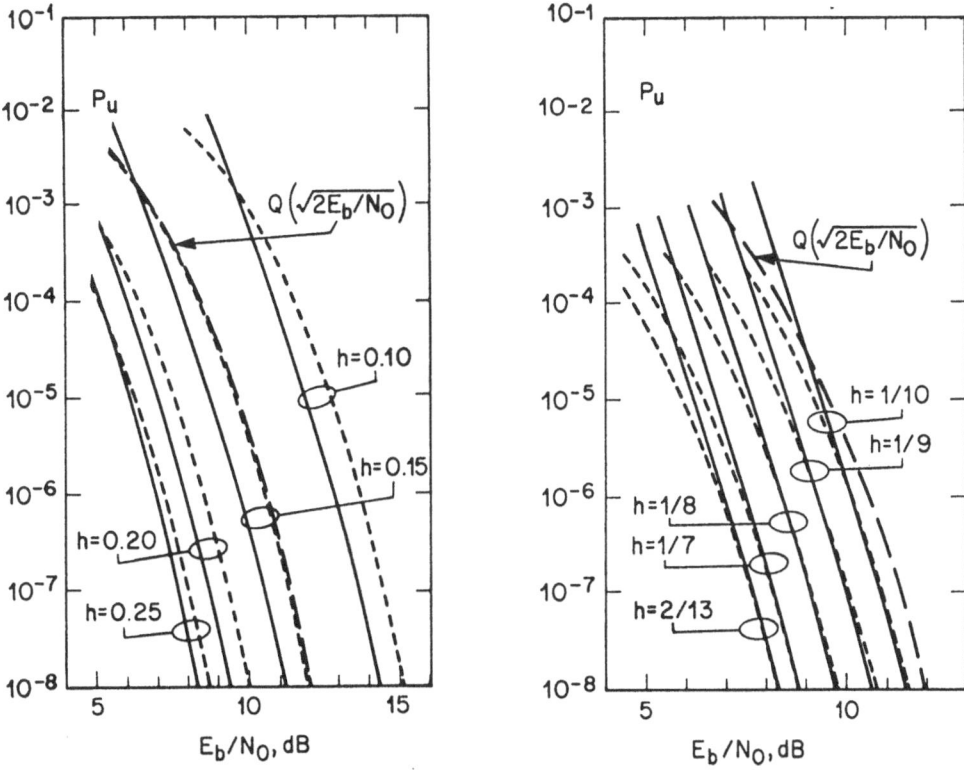

Figure 15. P_u (solid line) for the rate 1/2 (7,2) code with 4-level CPFSK.

Figure 16. P_u (solid line) for the rate 2/3 (15,2) code with 8-level CPFSK.

$C_{d_{\min}}$ can be calculated for the coded schemes with sufficient memory ν such that the encoder-independent upper bound on $d^2_{\min}$ is reached. This upper bound corresponds to error events with bit errors only in the most significant (uncoded) bit(s). It is found in [5] that, with rate 2/3 encoders and 8-level CPFSK, there are 8 minimum distance pairs for each start-state in the encoder. Furthermore, each such pair yields 2 bit errors and the error events have length $2T$. Thus, if the encoder-independent upper bound is reached, then $C_{d_{\min}} = 1/2$. A Viterbi receiver have been constructed in software [1],[14]. The variance of the estimated bit error probability depends on the number of error events simulated. For high bit error probabilities, 1000 bit errors have been used. For lower bit error probabilities fewer bit errors are used. The path memory of the Viterbi detector has in all the simulations been chosen to be 50 channel symbol intervals, which is large enough to ensure that all unmerged pairs have a distance greater than $d^2_{\min}$. Simulations have been done for some selected schemes, and these are given in [1],[15]. The simulated bit error probability is consistent with the calculated upper bound. It is interesting to note that the gains in E_b/N_0 for high signal-to-noise ratios based on the free Euclidean distance turn out to be pessimistic for many interesting combinations of convolutional codes and continuous phase modulation. We have seen examples of coding gains which are too low by up to 1 dB.

Throughout the work presented above we have used a scaled version of the uncoded power spectra for estimating the bandwidth of the coded CPM systems. This has been verified to be a very good estimate for good codes, see [16], [17]. Above, we have presented one approach to find good combined codes and CPM. Especially for MSK, but also for partial response schemes there is more detailed work on the integration of coding and modulation presented in [18], [19], [20]. A "simpler" way to obtain coded CPM is to use multi-h systems, [1], [21] as we mentioned above. A variation on this theme is the multi-T schemes where the symbol time varies cyclically with time and where h is kept fixed [22]. The results of multi-h and multi-T are similar. Coded CPM systems with convolutional codes are more efficient. References [23], [24], [25] and [26] give further overviews of developments in coded and uncoded CPM.

3. COMPARISONS TO TRELLIS-CODED QAM AND MPSK

Recent years have seen an avalanche of papers in the area of trellis-coded modulation [3]. Some of these results are summarized in the overviews [4], [23], [24] and in the books [1], [25]. Also see e.g. the papers in the special issues [26], [27] and the papers [28]-[35]. In this section we will briefly summarize these results for the purpose of comparing trellis-coded AMPM to coded CPM. We will use conventional signal constellations [4], [25]. All distance values used in this section is normalized with respect to the bit energy. Figure 17 illustrates the power and bandwidth tradeoff for trellis codes based on 2-dimensional signal sets for QAM, AMPM and PSK constellations. We have used a rule of thumb for the 99% power-in-band bandwidth, namely $2BT_b \approx 1.3/b$ where b is the number of bits/sec per Hz. Thus we allow a factor of 1.3 for realistic filtering. This seems reasonable studying the FCC mask, etc. The exact number of course depends on the actual filtering used. We will use this rule also in the next section when coded CPM is compared to trellis-coded QAM. We have also included some conventional convolutional codes combined with 4-PSK. These codes are taken from tables in [36]. Figure 17 shows the asymptotic gains (high SNR) based on free Euclidean distance for 4-state, 16-state and 256-state systems. The 4-state schemes yields a 3 dB improvement without bandwidth expansion. It takes about 256 states to get another 3 dB. The reference signal in Figure 17 is MSK (and 4-PSK) which have a normalized squared free Euclidean distance of 2. In Figure 17 we have only shown results for trellis-coded 2-dimensional signal sets. Further improvements can be obtained by utilizing a larger number of dimensions [4], [26]-[32]. The additional gains obtained are modest, typically not more than 1-2 dB. Further improvements are small and very difficult to achieve. This is not surprising, since for the Gaussian channel, the systems operate not far from (or at) the so called computational "cut-off" rate R_0 [4]. Further work on trellis-coded modulation should be devoted to issues such as interference limited channels [34],[35], fading channels and channels with imperfect carrier phase recovery, see e.g. [37]-[42].

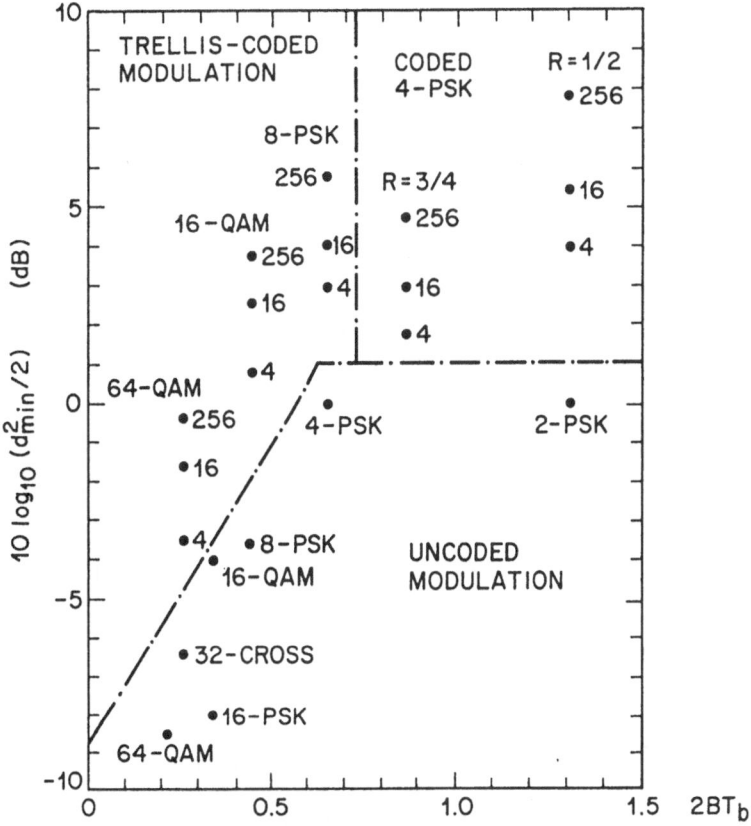

Figure 17. Power versus bandwidth for conventional coding of 4-PSK and for trellis-coded M-PSK and AMPM. The number of states is given for each system.

Above we have studied how system parameters such as c, m, A, h etc. influence the CPFSK system performance in terms of power and bandwidth efficiency. We will now also make an attempt to compare coded CPFSK to uncoded QAM, AMPM and combined convolutional codes and QAM (often referred to as trellis-coded QAM [4]). On purpose we ignore trellis-coded one-dimensional AM and trellis-coded M-PSK (except 8-PSK), since these systems are inferior to trellis-coded QAM. It is straightforward to evaluate the normalized minimum Euclidean distance of 4-QAM, 16-QAM, 64-QAM, 256-QAM and of 8-AMPM,

32-AMPM and 128-AMPM [4]. By 16-QAM we mean quadrature amplitude modulation in two dimensions with 4 levels in each dimension yielding 16 signal points. 8-AMPM is half of this constellation with improved minimum distance [3]. It is also straightforward to calculate the double sided normalized bandwidth (99% in band power). In our comparison we have, however, used a rule of thumb, namely the inverse of the number of bits/s/Hz of the modulation scheme increased by 30% [7], [43]. In Figure 18 we have plotted (as a reference) the power bandwidth tradeoff in terms of $10\log_{10}(d_{min}^2/2)$ versus $2BT_b$ for the best found uncoded CPFSK scheme, namely 8-CPFSK. We also show the results for the best known coded constructive schemes [1], [6], namely 16-CPFSK with $v = 4$ and the simpler $v = 2$. (v is the number of delay elements in the convolutional encoder.)

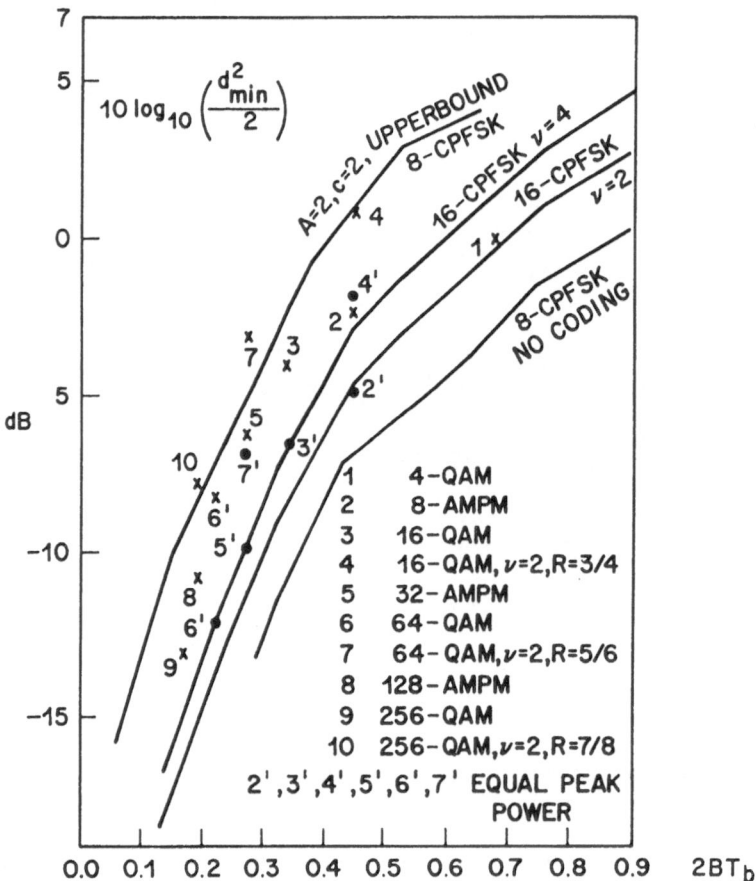

Figure 18. Power-bandwidth tradeoff for some selected coded CPFSK and trellis-coded QAM schemes.

Then we show the best upper bound for $A = 2$ schemes, which is 8-CPFSK with $c = 2$. For comparison, the QAM schemes are given as points (x) in this plot. Furthermore we also show some schemes with equal *peak* energy per information bit (o) rather than equal average energy per information bit (x). Some simple $v = 2$, trellis-coded QAM schemes are also shown in Figure 18, [7]. Note that the comparisons are given for a linear ideal channel. With this background it is not surprising that the QAM systems are better then the CPFSK schemes. After all, having the freedom of signal variation both in amplitude and in phase can be expected to yield better results than signals varying only in phase. Figure 18 indicates how large the differences are. From the results we conclude that the QAM systems are (not surprisingly) better then the uncoded CPFSK systems on a linear channel. The real issue comes with transmission over a nonlinear channel. We note that applying, say, 6 dB backoff [43] (i.e. reduce the output transmitter power because of nonlinearity) to some of the QAM schemes brings them very close to "simple" coded CPFSK or even uncoded 8-CPFSK. 6 dB is sometimes an optimistic backoff value [43]. Considerably larger values are sometimes required. However, more work has to be performed before it can be conclusively established whether coded CPFSK is preferred over coded QAM when transmitted over real moderately nonlinear channels. Clearly, it is going to depend on type of nonlinearity and filtering, predistorsion, backoff etc. Perhaps the engineering solution is a set of pseudo-constant amplitude signals suitably combined with coding.

The constant amplitude schemes have the potential of being transmitted with minimal loss over the highly nonlinear channel while the QAM systems will lose more. In this sense, Figure 18 is a baseline comparison with a maximal difference between the schemes. It should be pointed out that better trellis-coded QAM systems exist with an additional gain of say 2-3 dB [4] over the given simpler systems. The comparisons above are based on the minimum distance, i.e. difference in required E_b/N_0 at very high signal-to-noise ratios. The same comparison can also be repeated at, say, a bit error probability of 10^{-5}. The error coefficient for trellis-coded QAM systems [4] is typically much larger then that for CPFSK [1]. That will bring the curves somewhat closer together, typically not more than about 1 dB though. The complexity of the coded CPFSK system in Figure 9 and Figure 18 is also of interest. The number of states required in the decoder is upper bounded by the number of states in the convolutional code (2^v) times the number of phase states in the CPFSK system. The latter number is given by the (rational) modulation index value h. Note that the parameter A only influence the number of transitions in the trellis and not the number of states.

4. SUMMARY AND DISCUSSION

Above we have given a detailed overview of results on combined convolutional coding and continuous phase modulation. We have also given brief overviews of uncoded CPM and trellis-coded AMPM systems. More detailed overviews of these reas are available in the references. Most of the coded CPM results are given for 99% power-in-band bandwidth values in the range of 0.3-1.5. Thus we are considering a range of theoretical spectral efficiencies of the range 1-3 bits/s/Hz. Channel capacity indicates that an E_b/N_0 of 0 dB is required for error free transmission at 1 bit/s/Hz. At 2 bits/s/Hz, 2 dB is required and at 3 bits/s/Hz 4 dB is required. Table 1 summarizes some systems in the 1-2 bits/s/Hz category and Figures 17 and 18 gives wider summaries. From the results above we can conclude that constructive schemes exists to improve the power efficiency of 4-PSK by means of up to 6 dB in E_b/N_0 at unchanged bandwidth or improve the spectral efficiency by almost a factor of 2 without loss in power efficiency (see Figure 17). The same holds for constant amplitude modulations with MSK as a reference point, (see Table 1 and Figure 9). Improvements with other chosen reference points are of the same magnitude. Further improvement will probably not come easily. One reason for this is that we now operate about 5 dB from channel capacity at 10^{-6} for the trellis-coded AMPM systems. A more practical limit is the computational cut-off rate. Many claims that the best trellis-coded modulation systems now operate at this boundary.

In [1] and [2], extensive reference lists are given for works on continuous phase modulation (CPM) up to late 1985. In this section, we have primarily given some new references on continuous phase modulation with emphasis on combined convolutional codes and CPM. We also give a basic set of references on trellis-coded amplitude and phase modulation. Power and bandwidth efficient digital channel coding and modulation is a very active area of research. On the Gaussian channel we can probably not expect further significant improvements for trellis-coded modulation. Instead, other impairments need to be studied such as phase-jitter, fading, interference, etc. For constant amplitude modulation, the above impairments also need further study. Furthermore, nearly constant amplitude modulation methods should be studied further. There is still room for better and simpler CPM signal design for the Gaussian channel. A somewhat different class of coded CPM schemes is defined by the term multi-h codes [1]. In this case, the modulation index h is time-varying from symbol time to symbol time in typically a cyclic manner. Thus further memory is introduced in the transmitted signal and some improvements over the uncoded systems are to be expected. However this class of codes is somewhat limited and it is shown in [1], that combined convolutional coding and CPM clearly outperform the best known multi-h systems on the Gaussian channel. It is of course conceivable to combine coded CPM with the multi-h concept.

There still remain many theoretical problems with transmission by means of continuous phase modulation. We will mention two of them here excluding such problems as synchronization, transmission in interference and fading and analysis of the performance in the presence of realistic nonlinearities. The first is evaluation of the effect of filtering of CPM and the meaning of bandwidth of coded multilevel CPM systems. Some initial results with mild filtering are reported in [1]. By using distance calculations in the spectral domain, results under severe filtering were obtained in [44]. In the presentation above we largely ignored the problem of complexity. The number of states in "simple" trellis coded QAM systems with optimum receivers could be kept quite low (say 4 to 16) and yet a gain of 3-4.5 dB is obtained. With coded CPM it is somewhat different. The convolutional code memory is quite moderate for the studied schemes, while the number of modulator states (phase states for CPFSK and correlative states for partial response schemes) is very large especially for the interesting cases of low modulation index and a large number of modulator levels. Simplified near-optimum reduced-complexity receivers is therefore an important research topic. The so called M-algorithm (and modifications there of) which was initially conceived for sequential decoding of convolutional codes [45] have been applied to uncoded CPM [46], convolutional codes, trellis-codes AMPM modulation and reception in the presence of intersymbol interference [45]-[47]. By means of simulations it has been shown that the M-algorithm can operate with much fewer states than the optimum Viterbi algorithm for CPM and yet give near-optimum performance for many parameter combinations. The issue of computational cut-off rate and channel capacity of systems with constant amplitude and continuous phase restrictions also needs to be resolved. Some work is reported in [1], [48], [49]. Finally, the issue of implementations. It has not been the purpose of this chapter to address hardware or system application issues. However, coded CPM for high frequency, high speed (100's of Mbits/sec) satellite modems is considered in [50] and intermediate speed (100's of kbits/sec) modems for digital mobile radio applications in [51].

REFERENCES

[1] J. B. Anderson, T. Aulin and C-E. Sundberg, "Digital Phase Modulation," Plenum Publishing Co., New York, NY, 1986.

[2] C-E. Sundberg, "Continuous Phase Modulation," IEEE Communications Magazine, Vol. 24, No. 4, pp. 25-38, April 1986.

[3] G. Ungerboeck, "Channel Coding with Multilevel/Phase Signals," IEEE Transactions on Information Theory, Vol. IT-28, No. 1, pp. 55-67, January 1982.

[4] G. Ungerboeck, "Trellis-Coded Modulation with Redundant Signal Sets." Part I and Part II. IEEE Communications Magazine, Vol. 25, No. 2, pp. 5-21, February 1987.

[5] G. Lindell, "On Coded Continuous Phase Modulation," Dr. Techn. Thesis, Univ. of Lund, Lund, Sweden, May 1985.

[6] G. Lindell, C-E. Sundberg and A. Svensson, "Narrowband Coded Digital Modulation Schemes with Constant Amplitude," 1985 International Tirrenia, Italy, Sept. 1985. Publ. in "Digital Communications," North-Holland, 1986, pp. 59-69.

[7] G. Lindell and C-E. Sundberg, "Power and Bandwidth Efficiency of Coded CPM — A Comparison to Coded QAM," 1987 International Conf. on Communication Technology, Nanjing, China, November 1987. Conf. Proc., pp. 229-231.

[8] A. J. Viterbi, "Convolutional Codes and Their Performance in Communication Systems," IEEE Transactions on Communications Technology, Vol. COM-19, No. 5, pp. 751-772, October 1971.

[9] A. J. Viterbi and J. K. Omura, "Principles of Digital Communications and Coding," McGraw-Hill, 1979.

[10] J. K. Omura, "Performance Bounds for Viterbi Algorithms," International Conf. on Communications, ICC '81, Denver, Colorado. Conf. Rec., pp. 2.2.1-2.2.5.

[11] E. Biglieri, "High Level Modulation and Coding for Nonlinear Satellite Channels," IEEE Transactions on Communications, Vol. COM-33, No. 5, pp. 616-626, May 1984.

[12] E. Zehavi and J. K. Wolf, "On the Performance Evaluation of Trellis Codes," IEEE Transactions on Information Theory, Vol. IT-33, No. 2, pp. 196-202, March 1987.

[13] S. Benedetto, M. Ajmone Marsan, G. Albertengo and E. Giachin, "Combined Coding and Modulation: Theory and Applications," IEEE Transactions on Information Theory, Vol. IT-34, No. 2, pp. 223-236, March 1988.

[14] G. Lindell, C-E. Sundberg and A. Svensson, "Bit Error Probability of Coded CPM-Bounds and Simulations," GLOBECOM '85, New Orleans, Dec. 1985, Conf. Rec., pp. 22.3.1-22.3.7.

[15] G. Lindell and C-E. Sundberg, "An Upper Bound on the Bit Error Probability of Combined Convolutional Coding and Continuous Phase Modulation," IEEE Transactions on Information Theory, Vol. IT-34, No. 5, pp. 1263-1269, September 1988.

[16] P. K. M. Ho and P. J. McLane, "The Power Spectral Density of Digital Continuous Phase Modulation with Correlated Data Symbols" Part One and Part Two. IEE Proc., Vol. 133, PtF, No. 1, pp. 95-114, February 1986.

[17] P. K. M. Ho and P. J. McLane, "Spectrum, Distance and Receiver Complexity of Encoded Continuous Phase Modulation," IEEE Transactions on Information Theory, Vol. IT-34, No. 5, pp. 1021-1032.

[18] F. Morales-Moreno and S. Pasupathy, "Structure, Optimization, and Realization of FFSK Trellis Codes," IEEE Transactions on Information Theory, Vol. IT-34, No. 4, pp. 730-751, July 1984.

[19] F. Morales-Moreno and S. Pasupathy, "Matched Encoders for Partial-Response Channels," Technical Report, EE, University of Toronoto, Canada, May 1987.

[20] B. E. Rimoldi, "Continuous Phase Modulation and Coding for Bandwidth and Energy Efficiency," Doctor of Science Thesis Diss. ETH No. 8629, Zürich, Switzerland 1988.

[21] J. B. Anderson and D. P. Taylor, "A Bandwidth – Efficient Class of Signal Space Codes," IEEE Transactions on Information Theory, Vol. IT-24, No. 6, pp. 703-712, November 1978.

[22] P. Szulakiewicz and W. Holubowicz, "M-ary Multi-T Phase Coders," IEEE International Symposium on Information Theory, Ann Arbor, Michigan, October 1986. Book of Abstracts, pp. 88-89.

[23] S. G. Wilson, "Bandwidth – Efficient Modulation and Coding: A Survey of Recent Results," ICC '86, Toronto, June 1986. Conf. Rec., pp. 965-969.

[24] C-E. Sundberg, "Theoretical Aspects of Achieving Power and Bandwidth Efficiency," Phoenix Conference on Computers and Communications, Phoenix, Arizona, March 1987. Conf. Proc., pp. 601-602.

[25] S. Benedetto, E. Biglieri and V. Castellani, "Digital Transmission Theory," Prentice Hall, NY, 1987.

[26] Special Issues on Bandwidth and Power Efficient Coded Modulation, IEEE Journal on Selected Areas in Communications, Vol. 8, August 1989 and December 1989.

[27] Special Issue on Voiceband Telephone Data Transmission, IEEE Journal on Selected Areas in Communications, Vol. SAC-2, No. 5, September 1984.

[28] G. D. Forney, R. G. Gallager, G. R. Lang, F. M. Longstaff and S. U. Qureshi, "Efficient Modulation for Bandlimited Channels," IEEE Journal on Selected Areas in Communications, Vol. SAC-2, No. 5, pp. 632-647, September 1984.

[29] L.-F. Wei, "Rotationally Invariant Convolutional Channel Coding with Expanded Signal Space, Part I: 180° and Part II: Nonlinear Coding," IEEE Journal on Selected Areas in Communications, Vol. SAC-2, No. 5, pp. 659-686, September 1984.

[30] S. G. Wilson, H. A. Sleeper and N. K. Srinath, "Four-Dimensional Modulation and Coding: An Alternate to Frequency-Reuse," ICC '84, Amsterdam, May 1984, Conf. Rec., pp. 919-923.

[31] L.-F. Wei, "Trellis-Coded Modulation with Multidimensional Constellations," IEEE Transactions on Information Theory, Vol. IT-33, No. 4, pp. 483-501, July 1987.

[32] G. D. Forney, Jr., "Coset Codes I: Geometry and Classification" and "Coset Codes II: Binary Lattices and Related Codes," In submission to the IEEE Transactions on Information Theory.

[33] G. Ungerboeck, J. Hagenauer and T. Abdel-Nabi, "Coded 8PSK Experimental Modem for the Intelsat SCPC-System," 7th International Conference on Digital Satellite Communications. München, Germany, May 1986, Conf. Proc., pp. 299-304.

[34] M. Kavehrad, P. J. McLane and C-E. Sundberg, "On the Performance of Combined Quadrature Amplitude Modulation and Convolutional Codes for Cross-Coupled Multidimensional Channels," IEEE Transactions on Communications, Vol. 34, No. 11, December 1986.

[35] M. Kavehrad and C-E. Sundberg, "Bit Error Probability of Trellis-Coded Quadrature Amplitude Modulation Over Cross-Coupled Multidimensional Channels," IEEE Transactions on Communications, Vol. COM-35, No. 4, pp. 369-381, April 1987.

[36] G. C. Clark and J. B. Cain, "Error-Correcting Coding for Digital Communications," Plenum Press, New York, 1981.

[37] M. K. Simon and D. Divsalar, "The Performance of Trellis Coded Multilevel DPSK on a Fading Mobile Satellite Channel," International Conference on Communications, ICC '87, Seattle Washington, June 1987, Conference Record, pp. 732-738.

[38] D. Divsalar and M. K. Simon, "The Design of Trellis Codes for Fading Channels," Presented at IEEE Communication Theory Workshop, Howey-in-the-Hills, Florida, April 1987.

[39] D. Divsalar and M. K. Simon, "Trellis Coded Modulation for 4800-9600 bits/s Transmission over a Fading Mobile Satellite Channel," IEEE Journal on Selected Areas in Communications," Vol. SAC-5, No. 2, pp. 162-175, February 1987.

[40] S. G. Wilson and Y. S. Leung, "Trellis-Coded Phase Modulation on Rayleigh Channels," International Conference on Communications, ICC '87, Seattle Washington, June 1987, Conf. Rec., pp. 739-743.

[41] P. J. McLane, P. H. Wittke, P. K.-M. Ho and C. Loo, "PSK and DPSK Trellis Codes for Fast Fading, Shadowed Mobile Satellite Communications Channels," International Conference on Communications, ICC '87, Seattle, June 1987, Conf. Rec., pp. 726-731.

[42] J. Hagenauer and C-E. Sundberg, "On the Performance Evaluation of Trellis-Coded 8-PSK with Phase Offset," International Conference on Communications, ICC '88, Philadelphia, Pa., June 1988, Conf. Rec., pp. 23.4.1-23.4.7.

[43] H. Yamamoto, K. Kohiyama, O. Kurita, M. H. Meyers, V. K. Prabhu, G. Hart and J. A. Steinkamp, "Future Trends in Digital Radio," IEEE Communications Magazine, Vol. 25, No. 2, pp. 40-52, February 1987.

[44] N. Seshadri, "Error Performance of Trellis Modulation Codes on Channels with Severe Intersymbol Interference," Ph.D. Thesis, ECSE Dept., Rensselaer Polytechnic Institute, New York, October 1986. Also see N. Seshadri and J. B. Anderson, "Asymptotic Error Performance of Modulation Codes in the Presence of Severe Intersymbol Interference," IEEE Transactions on Information Theory, Vol. IT-34, pp. 1203-1216, September 1988.

[45] C.-F. Lin and J. B. Anderson, "M-Algorithm Decoding with Path Recovery for Convolutional Channel Codes," IEEE Global Telecommunications Conference, GLOBECOM '87, Houston, Texas, December 1986, Conf. Rec., pp. 181-185.

[46] T. Aulin, "A Fractional Viterbi-Type Trellis Decoding Algorithm," IEEE International Symposium on Information Theory, Ann Arbor, Michigan, October 1986, Book of Abstracts, p. 141.

[47] S. J. Simmons, "Breadth-first Trellis Decoding with Adaptive Effort," To appear in IEEE Transactions on Communications.

[48] I. Bar-David and S. Shamai (Shitz), "On Information Transfer by Envelope-Constrained Signals over the AWGN Channel," IEEE Transactions on Information Theory, Vol. IT-34, No. 3, pp. 371-379, May 1988.

[49] S. Shitz and I. Bar-David, "Capacity Bandwidth Trade-off for a Class of Constant Envelope Modulations," IEEE International Symposium on Information Theory, Brignton, England, June 1985, Book of Abstracts, pp. 100-101.

[50] C. Ryan, "Hardware Realization of Bandwidth/Power Efficient Modems," Sixth Annual International Phoenix Conference on Computers and Communications, Phoenix, Arizona, February 1987, Conf. Proc., pp. 145-147.

[51] T. Maseng and O. Trandem, "Adaptive Digital Phase Modulation," Second Nordic Seminar on Digital Land Mobile Radio Communication, Conf. Proc., pp. 64-69, Stockholm, Sweden, October 1986.

Lecture Notes in Control and Information Sciences

Edited by M. Thoma and A. Wyner

Lecture Notes in Control and Information Sciences

Edited by M. Thoma and A. Wyner

Lecture Notes in Control and Information Sciences

Edited by M. Thoma and A. Wyner